AF256021

"INSIGHTS"

THE NATURAL GAS INDUSTRY OF TRINIDAD AND TOBAGO.

IN TRANSITION?

By
René Monteil

ISBN 978-1-916540-36-1
First Revised Edition 2023

Contents

Ever remembering that Nature has implanted in each person's breast a sacred and indissoluble attachment toward that country whence he has derived his birth and infant nurture

Acknowledgments

A picture is worth a thousand words!

My deepest gratitude to the pioneers of the local natural gas industry whose inspirational leadership at Trintoc, the NEC, and NGC has indelibly stamped within me a blueprint for pursuing the continued development of the energy sector in Trinidad and Tobago.

I thank all of my former colleagues who have assisted me throughout my professional career, particularly Martin Houston, Gordon Shearer, and Simon Bonini, for their personal support and the 'fun' we had during the Atlantic LNG adventure.

Finally, I express my heartfelt and profound appreciation of the diligent efforts by various persons who have provided critical inputs in the production of this book:

(a) for the newspaper reports, Anton Cooke, the librarian at Guardian Media, and Avril Belfon, the archivist, and her staff at the National Archives of Trinidad and Tobago; and

(b) for the material that I had written, the meticulous technical oversight by Eugene Tiah, a long-standing colleague in the energy sector, the quality control by Pauline Philip, copy editor, and ultimately, my daughter, Dominique, for the eventual advice on publication.

About the Author

I have a few vivid memories of my time in primary school. At Rosary Boys R.C. in stage 3, after I passed the examination for the First Communion, it was discovered that I was six years old, which caused turmoil among the nuns who did not know what to do. Ultimately, I made my 1st Communion with the consent of the priest, who simply declared that I had attained "the age of reason." In Standard 5, I was distraught when my classmates were allowed to sit the Common Entrance Examination, and I had to be held back for a 2nd year because I was ten years old. After pleading with my mother, she took me out of the school, and I sat the exam under the tutelage of Mr. St Elmo Gopaul at Morvant E.C. (which was co-educational). At the time, my mother was a District nurse based in Morvant, and the school was a stone's throw from the health centre. I remember every morning, at the start of class, we turned to the East and said the Apostolic Creed. On the day of registration at St Mary's College, my mother complained to a priest that I had claimed the $24 scholarship money as a personal prize.

My entry into form 1 in St. Mary's College was exciting but not daunting because, when we were still in primary school, my father would take my older brother André and me to annual 'Intercol' matches in the Queen's Park Savanah between, at that time, QRC and CIC. My father had entered St Mary's college in 1923, thereby following the footsteps of my grandfather and my great-grandfather, Alfred Monteil. My great-uncle, "Uncle Freddie," the youngest son of Alfred Monteil, was the first local member of the "Holy Ghost Fathers" (C.S. Sp.). Uncle Freddie had entered the priesthood at 19 years of age, was ordained in 1925, and died in 1975, a few months short of his golden anniversary as a priest. He had spent most of his life as a missionary in Haiti.

I have no fear of vaccination. As a county health visitor, my mother had inoculated us against every disease that hit the region, including South America. We were also given "Seven Seas" cod liver oil daily, and to this day, I still either chew the capsules or 'guzzle' it from a bottle. My mother also took the lead as a parent when my brothers and I needed serious medical care because of injuries. André had had an appendectomy before the age of 5; fell upon a spike that cut into his forehead; broke his right arm in two places and, later, his left arm in two places; and was knocked down by a car and hospitalised. My younger brother, Roger, broke only one arm but in two places. I was relatively unscathed because I received only one fracture in my left arm when I fell off a tree, playing Tarzan with my cousin Brian in Victoria Square. The impact on my mother of our several accidents was exacerbated because of my father's aversion to the sight of blood. When I had ten stitches in the palm of my left hand, after it was split open by a galvanised sheet (while jumping a fence to retrieve a 'Graham' mango at Alvin ("Micky") Fitzpatrick's home), a nurse had to revive my father with smelling salts while he remained in the car outside the hospital.

As a family tradition, we were taught to play the piano in our 6th year, and, initially, André and I had music lessons twice a week. However, about two months before my grade 2 Royal Academy of Music practical piano examination, our music teacher returned from an extended holiday in England to discover that, in her absence, both André and I had done little preparation for our respective examinations. Consequently, she decided that it was imperative for us to have piano lessons on weekdays from 5.00 am until 6.30 a.m., which led to André achieving a merit in the grade 3 examination. I attained a distinction in grade 2. Thereafter, my father woke us up at 4.45 a.m. (aided and abetted by a clock that chimed quarterly). Even if the clock had stopped (because it had not been wound) or if, on the rare occasion, my father had overslept, we did not escape because the teacher would ring for us at 5.00 a.m. Music lessons ended when I was in lower form 6 when I passed the grade 8 theory exam (having passed the grade 8 practical exam the year before). Today, I still get up routinely before 4.30 a.m. and fall asleep long before 8.00 p.m.

Having come from a very sporting family, I had learnt to leg spin with a 'cork-ball' before I went to St Mary's College. We usually played cricket in the Sacred Heart Church schoolyard after 8.30 a.m. mass on Sundays. On public holidays, an alternative venue was the Fitzpatrick's driveway, which, regrettably, restricted on-side shots for right-handers because of a rose garden. And, as a teenager, we also used to play football in Victoria Square until around 5.45 p.m. when the City Police would chase us for contravening the well-displayed prohibition against "football or rounders." In forms 4 and 5, I became a permanent fixture on the A Colts cricket team and, thereafter, represented the college on the 1st XI team for three years. A consequence was that I obtained very poor grades in my 6 O' levels on my first attempt, but the following year, I received the requisite distinctions that permitted me entry to Lower 6 Science. On mature reflection, I believe that those competitive sporting activities had prepared me for the checkerboard of life by extolling the virtues of teamwork, discipline, hard work, and perseverance.

It is very likely that my career choices were influenced by my father, and from my early childhood, I remember him mixing potions. He gargled every morning with a purple solution that he called "pot permang." Later, I learnt that "pot permang" was potassium permanganate, which he would obtain from a drug store at the corner of Duke and Charlotte Streets in Port-of-Spain. As a firm believer in naturally occurring medicines and a keen sportsman, my father would rub himself with "Oil of Wintergreen" whenever he suffered a muscular injury as an anti-inflammatory and topical pain reliever. When I studied organic chemistry in form 6, I learnt that Oil of Wintergreen is the methyl ester of salicylic acid, and it is closely related to aspirin. Perhaps it is no coincidence that, at the age of 5, I received my first chemistry set. Later, as a teenager, I was able to kindle my own, perhaps more adventurous, interest in chemistry by purchasing materials from the same drugstore that my father had used. Potassium chlorate, when ignited on its own, became my 'rocket' propellant or, with sulphur and charcoal, produced gunpowder for a New Year's bang.

After my father retired at the age of 65, he took up legal studies as an external law student at Gray's Inn, London. In those days, one could enroll at one of the Inns of Court, take the examinations in the subjects individually, and ultimately qualify as a barrister after passing all the examinations. Although he had abandoned his studies after passing criminal law, his skirmishes with the law did influence his record-keeping and decision-making, which relied heavily on evidence.

As regards character development, the 1950s and 1960s were truly my golden years as during that time, I received my primary and secondary education and critical social development. Rosary Boys R.C. primary school, in combination with my parents, had inculcated the tenets of morality and respect for others. "Who made you?" "Why did God make you?" I can still recite and understand the eight Beatitudes, the four Cardinal virtues, and the seven virtues and their contrary vices. Later, during my secondary education, St Mary's College put more flesh on the bones after the die had been cast and the foundation stone laid at primary school.

1970 was a time of social upheaval in Trinidad and Tobago, and in Form 6 Science A, as we were about to embark upon our future careers, Fr. Gerry Pantin, the spiritual leader of our class, had emphasised and, by his example, demonstrated, *"To whom much is given, much is expected."* My A' level results were not flattering as I had barely passed Mathematics (Pure & Applied) and Physics, although I did obtain an A-grade in Chemistry. Before leaving for England, my father taped David (Toby) and me about our career aspirations. David said that he was going to Jamaica to become a doctor, and I confirmed that I would be going to England to do a PhD in organic chemistry. In October 1970, I left Trinidad, and with the confidence instilled within and a college scholarship after my 1st degree, my objective was achieved in January 1977. Thereafter, I undertook post-doctoral research fellowships in England until my appointment as a lecturer in Chemistry at the University of the West Indies in Mona, Jamaica. My achievements, as a chemist, were chronological events in a chain reaction that had been triggered when I was five years old.

I have often been asked: "Why did you switch to Law?" The answer is simple. It was expedient to do so after my wife, Michele, had achieved both the pre-clinical and clinical medals in medical school in Jamaica. Thus, we knew that, at some stage, the family would have to return to England so that she could complete her post-graduate education in medicine. So, what would I do? After considering the options and my long-term projections, the law seemed to be the best choice.

I began legal studies in England in October 1985 when Michele started the second year of her 2-year internship at the Port-of-Spain General Hospital. The timing was feasible because of the ages of our two girls. Noele, then five, who was nine months old when Michele started medical school, and Dominique, then three, was born when Michele was in her 2nd pre-clinical year. We always had strong support from the extended family. Michele and the two girls joined me 12 months later at my mother's home in Croydon, Surrey, while I was a student at the Council of Legal Education in London preparing for the Bar examination. I was called to the Bar of Middle Temple in July 1987. Later, during pupillage in April 1988, Danielle, our 3rd daughter, greeted the world loudly on the 2nd 'bong' of Big Ben at 10.00 p.m. at the Westminster Hospital, where Michele was specialising in clinical immunology.

I was admitted to practise in Trinidad and Tobago in February 2001, and Micky, then a senior counsel, was one of my sponsors. This milestone was achieved in a circuitous manner. When I started my legal studies, any Barrister who had qualified at the English Bar was entitled to be admitted to practise in Trinidad & Tobago. However, while at the Inns of Court in 1986, legislation was enacted in Trinidad, which removed the entitlement, subject to a transitional provision that applied only to those who had commenced legal studies <u>before</u> 1st January 1985. Thus, at face value, the legislation took away my right unfairly because I had started legal studies before the legislation was enacted with a legitimate expectation that I would be admitted to practice after successfully completing my studies.

In August 1987, I completed the Petition with other relevant documents to be admitted to practice, but the restriction in the transitional provision was discovered a few moments prior to filing the documents in the High Court. My disappointment was exacerbated when persons who had started legal studies on or before 1st January 1985 and had passed the Bar exam with me in 1987 (or even after me in 1988), were admitted to practice. Eventually, as a result of policy changes, after almost 15 years, I was admitted as an attorney in the High Court in Trinidad.

My most exciting professional experiences occurred during my PhD studies with Professor Gerhard V Boyd, with whom I had great fun when synthesising several novel heterocyclic compounds (https://pubs.rsc.org/en/content/articlelanding/1980/p1/p19800000846). My time spent in the laboratory was enjoyable, often unsuccessful, but ultimately rewarding when I 'discovered' the "Acyl Analogues of the Ene Reaction." It was a momentous occasion that provided yet another life lesson about the benefits of discipline, hard work, and perseverance. Gerhard, German-Jewish, had fled Germany at 16 years of age. He gave 2nd-year undergraduate organic chemistry lectures in his lab coat, and he spoke English with a strong German accent. I was very impressionable. Often, when he spoke, his accent reminded me of Fr. Ivan Galt's repeated injunction during A' level chemistry classes about the "Wurtz Synthesis": "There is no 'W' in German; it is a double V."

Professor Ken Julien was also influential in my working life, and I got to know him very well after many years of public service. We first met in 1984 when he chaired an NEC committee that recruited me to the company. Many years later, after a meeting with him in London with Malcolm Jones, I took up an appointment with NGC as the LNG Coordinator for the Atlantic LNG Train 1 Project. Professor Julien's work output was breathtaking.

As a university undergraduate, I started a search for Truth and, in my quest, I went to different places of worship. I never found an answer in any single religion but I eventually concluded that many religious beliefs are consistent with scientific principles and the principle of the balance: Nature restores equilibrium (Le Chatelier, a chemist) and to every action, there will be an equal and opposite reaction (Newton, a physicist). Newton had discovered what the priests in Egypt had kept hidden in their sanctuaries – the Scales of Justice of the Old Covenant – "whoso sheddeth Man's blood, by Man, shall his blood be shed." And, does "..whatsoever a man soweth, that shall he also reap " (Galatians 6:7) and the concept of Karma in Hinduism and Buddhism, have the same root, the law of cause and effect? In local parlance, "what goes around, comes around."

Thus, today, I try to keep an open mind about different religious beliefs because perspectives of Truth are likely to vary when viewed through lenses that are formed by different social and cultural experiences.

Ultimately, I readily accept that life is dynamic, variables keep changing, and generations that follow me might well have to solve more complex equations than those that were given to me.

RLM

November 2022

Preface

At the outset, Insights reviews Trinidad and Tobago's pedigree as one of the oldest petroleum provinces in the world. Historically, the Nation had attracted many global players to invest in projects across the entire value chain of the energy sector. There was exploration and production activity of Trinidad and Tobago's hydrocarbons since 1857 and by 1926, Trinidad and Tobago had become the largest oil producer in the British Empire. In the early 1900s, many local refineries were established including a refinery owned in 1912 by the United British Oilfields of Trinidad ("UBOT", later Shell) and in 1917, by Texaco. With the discovery of large volumes of natural gas and declining oil production, the country's first Prime Minister, Dr. Eric Williams, set out a Vision for the transformation of the local economy from one based upon oil to one based upon natural gas.

As a result of policy changes and direct foreign investment in the local natural gas sector, Trinidad and Tobago became a major global natural gas development centre. Atlantic LNG Train 1 was the 1ˢᵗ LNG plant to be constructed in the Western Hemisphere since 1969. The 4ᵗʰ LNG Train, was, at the time of its construction, the largest in the world and the Atlantic LNG plants reflected one of the largest worldwide liquefaction facilities. The Caribbean Nation became the largest supplier of LNG to the United States and currently supplies LNG to global destinations, including Spain, where LNG supplies have provided security of gas supply. The Nation also has worldwide recognition among the major stakeholders in petrochemicals (ammonia and methanol) and, at its peak, one of the methanol producers had the largest global methanol production. Thus, Trinidad and Tobago became a global player in the natural gas industry.

In Chapter 2, my introduction to the energy sector, I explain crude oil processing and describe how indigenous raw materials such as crude oil and natural gas can be used to produce more valued-added products (i.e., "petro" – "chemicals") that are important intermediates in the chemical industry. The production of such value-added products was a strategic objective of the NEC. Some of my more memorable experiences as a legal advisor in the Department of Energy in Britain are identified in chapter 3, which include a cross-border interconnector gas pipeline treaty and highly controversial environmental issues that arose during North Sea operations. These technical and legal experiences provided the foundation for my later leadership roles in the natural gas sector.

A core feature of "Insights" is that it provides (hitherto unknown) contemporaneous analyses of the Atlantic LNG commercial negotiations that transpired during the development of the four Atlantic LNG Trains and the attendant overlaps of shareholder interests that brought the shareholders to the brink of litigation. Going beyond the benefits of the revenue from LNG production, I have questioned whether the commercial tolling structures that were approved for LNG Expansions have allowed gas producers to control the marketing of LNG with suboptimal benefits for Trinidad and Tobago. Atlantic 1, the Train 1 entity in which NGC, the State-owned gas company was a shareholder, had provided an intermediate institutional framework to prevent complete vertical integration by the gas producers.

*The Train 1 merchant structure had also provided an opportunity to audit the revenues obtained from the marketing arrangements to Boston and Spain. However, for LNG Expansions, the tolling arrangements (which were first introduced with the Train 2-3 project, a watershed) had allowed the diversion of LNG supplies to global markets at very high prices without any share of the diversion upside being available to Atlantic LNG, which led to significant offshore value loss for the Government. The situation was complicated by the Government not having audit rights in respect of **<u>all</u>** aspects of the business of the marketing affiliates of the gas producers, a limitation that exacerbated the challenges of determining the appropriate netbacks to the wellhead.*

Thus, insofar as adjustments have not been made over the last 20 years, in the final chapter, which considers the future challenges, Insights has recommended that the Government use the Heads of Agreement with Shell and bp to obtain audit rights and greater control of the choice of LNG markets. It might also be an opportunity for the Government to adopt principles to control how the Nation's natural gas resources are utilised in the future. The control of Trinidad and Tobago's hydrocarbon resources has now become a strategic imperative because of the geopolitical consequences of the war in Ukraine, which is continuously changing the global supply/demand energy balance. Moreover, decisions about the availability or utilisation of Trinidad and Tobago's resources, which have political implications, would be better left directly within the purview of the

State and not with transnational companies with external global interests that are not necessarily aligned with those of Trinidad and Tobago.

Some of the more controversial issues in the local energy sector are also tackled head-on. Notwithstanding global trends towards liberalisation of energy sectors, Insights explains why NGC's status quo should be maintained for its merchant roles but cautions that NGC should avoid conflicts of interest by investing in the same markets as its customers. Chapters are devoted to strategic value-added projects for diversifying the uses of natural gas, such as an Aluminium smelter and an ethylene complex utilising ethane from indigenous gas streams. Readers have an opportunity to assess the quality of the decision-making that led to the cancellation of the Alutrint project and the failure, so far, to implement an ethylene project.

The final chapter recognises that the sins of the past, committed twenty or so years ago by the Government, like chickens, have now come home to roost. As an illustration, in order to preserve the longevity of the energy sector in an increasingly competitive global environment, the Government now has to rely upon the unpredictable results of Bid Rounds and the geopolitical challenges of implementing the strategic cross-border initiatives with Venezuela (which are now promising due to U.S. policy changes).

Based upon the Trinidad and Tobago experience in the energy sector, "Insights" ultimately provides two key lessons for Third World countries as they implement policies to develop their depleting natural resources. Firstly, the historical importance of State control of its critical resources and its relationship to a country's destiny is well documented. This conclusion formed an important aspect of Dr Williams' Vision for the energy sector, which is borne out when he announced the formation of the State-owned oil company, TRINTOC, on Independence Day in 1974:

> *"The real question was not whether the oil flowed ..... but to whom did the benefits flow – U.B.O.T. ..... United British first, Trinidad last. Then the name was changed ...... Shell first, Trinidad last .... Reverse that, put Trinidad and Tobago first – TRINTOC. Now we know not only from where the oil flows, but to whom the benefits flow."*

However, today, as one now reflects upon the status of the Trinidad and Tobago energy sector, the following irony cannot be escaped: first, Shell, BP and Texaco owned Trinidad and Tobago's oil assets; after they left these shores for economic reasons, the upstream and downstream interests were vested in Petrotrin and TTNPMC, both State entities; and now, with the transformation from oil to natural gas, through vertical integration and control of the LNG value chain, these companies control a critical engine for developing the Trinidad and Tobago economy. As early as January 1976, Dr Williams had said:

> *"There have been attempts to persuade us that the simplest and easiest thing to do would be to sit back, export our oil, export our gas, do nothing else and just receive the revenues derived from such exports and as it were, lead a life of luxury – at least for some limited period.... ."*

Has Trinidad and Tobago become economically re-colonised? Plus ça change, plus c'est la même chose?

As a second observation for Third World countries, Insights notes that transnational companies, like all private corporate entities, generally aim to maximise their shareholder value as quickly as possible. However, rapid exploitation of a depleting natural resource, without other considerations, might not facilitate sustainable long-term economic benefits for a host country. He who pays the piper often calls the tune particularly when a country is in financial difficulties. However, if the 'tune' is not in harmony with the best long-term interests for the host country, which might not be easily discerned if there is a lack of transparency in decision-making, there can be even more severe economic consequences for future generations of the citizens of the country, culminating in a failed State. With an injunction to policy makers in the Third World, Insights cautions that those who fail to learn from history are doomed to repeat it.

Finally, Insights considers the Paris Agreement, which came into force on 4th November 2016 and is the principal regulatory instrument governing the global response to climate change. As regards steam reforming of natural gas, Insights explains why carbon capture utilisation and storage, to produce "blue" hydrogen, is highly debatable as an expedient to reduce greenhouse gas emissions and a cost-effective substitute for "green" hydrogen. Ultimately, the biggest challenge might be the cost of re-making industries that currently lie at the center of the economy and our lives – using green hydrogen as a substitute for all existing grey hydrogen to make ammonia (which is globally required for fertilizer production), would have a severe knock-on effect on food prices.

Common Terminology

Units. Measurement is a cornerstone of trade, science, technology, and quantitative research in many disciplines. The **International** System of Units (SI), commonly known as the metric system, is the international standard for measurement. However, before SI units were widely adopted worldwide, the British systems of English units, later imperial units, were used in Britain, the Commonwealth, and the United States. The petroleum industry still widely uses imperial units.

	METRIC	IMPERIAL
Length	millimetre, centimetre, metre, kilometre	inch, foot, yard, mile
Mass	milligram, gram, kilogram	ounce, pound, stone
Capacity	millilitre, centilitre, litre	pint, gallon

The Context. The choice of units can vary in different contexts. For example, we refer to an 8-furlong horse race, but an athlete runs a mile (nowadays, a metric mile) in a stadium.

Depending on the purpose of the measurement, the products from the oil and gas industries may be measured in different ways. Petroleum engineers, particularly those working in the Western Hemisphere, measure <u>volumes </u>to answer: "How much oil or gas?" On the other hand, ship owners might prefer oil shipments to be measured by <u>weight </u>to avoid overloading their tankers.

Volume. Volume reflects capacity. Thus, the unit of volume is a unit for measuring the capacity or the extent of an object or space. The SI unit of volume is the cubic metre (m^3), but the volume of natural gas that flows through a pipeline can also be measured in cubic feet (imperial units).

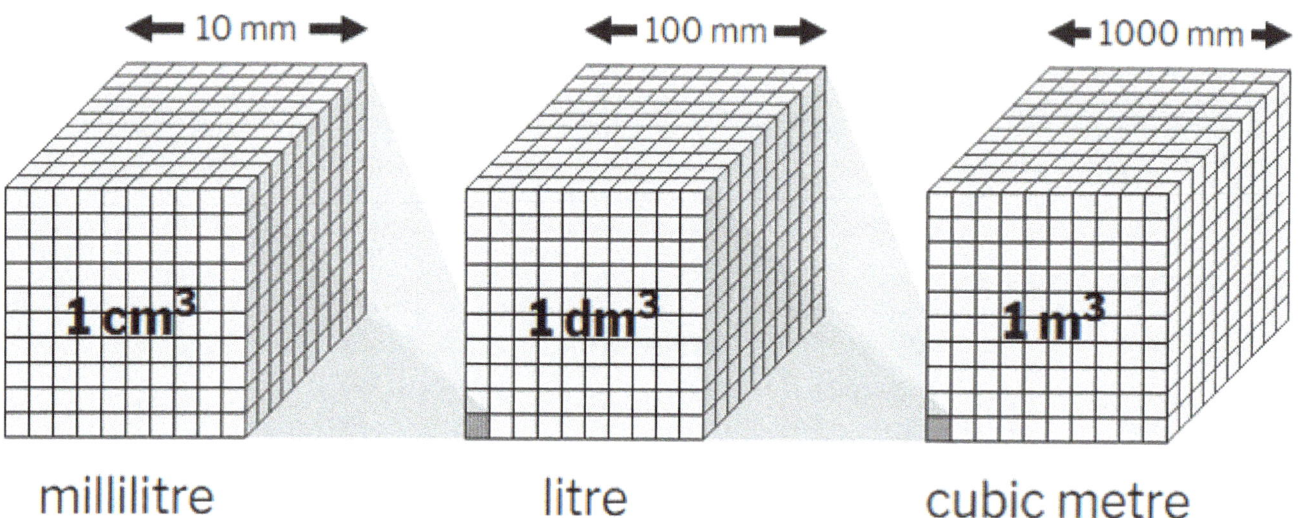

millilitre   litre   cubic metre

The volume of gas in a container varies significantly with changes in temperature and pressure because gas is compressible. And in order for gas volume measurements to have any comparative significance, they must have a standard frame of reference. Thus, the industry has established standard conditions when referring to gas volumes. When calibrated under standard conditions,[1] in the imperial system, which is used in the U.S., the unit is the "standard cubic foot" (scf).

Metric Tonnes. In Europe, where metric units are generally used, crude oil is measured in terms of weight and expressed in metric tonnes (MT). One MT is equal to 1000 kilograms (2,204 lb.)

---

[1] A majority of the natural gas industry in North America have adopted 60 °F and 14.73 psia as their standard reference conditions for expressing natural gas volumes and flow rates.

The imperial unit, the ton, is equal to 2,240 lb. In the oil industry, barrels (bbls) are still used as the measure of the volume of oil in the U.S. One barrel is equivalent to 42 U.S. gallons.

Depending upon the context, metric tonnes, cubic metres (m^3), or cubic feet are used to measure quantities. For example, the quantity of the gaseous feedstock that is supplied to a petrochemical plant is often described in terms of the volume of cubic feet, whereas the amount of the material that is produced after manufacturing is often expressed in terms of metric tonnes.

Hydrocarbons. Hydrocarbons are a class of organic chemicals that are made up of only the elements of carbon (C) and hydrogen (H). They are the principal constituents of crude oil and natural gas. Crude oil is a very complex mixture that contains thousands of hydrocarbons. The simplest hydrocarbon, methane (CH_4), is the main constituent of natural gas, which contains hydrocarbons of increasing chain length, chemically similar to methane, called "alkanes".

ALKANES

IUPAC naming system:

Molecular formula	Condensed Structural Formula	Name
CH_4	CH_4	**meth**ane
C_2H_6	CH_3CH_3	**eth**ane
C_3H_8	$CH_3CH_2CH_3$	**prop**ane
C_4H_{10}	$CH_3CH_2CH_2CH_3$	**but**ane
C_5H_{12}	$CH_3CH_2CH_2CH_2CH_3$	**pent**ane
C_6H_{14}	$CH_3CH_2CH_2CH_2CH_2CH_3$	**hex**ane
C_7H_{16}	$CH_3CH_2CH_2CH_2CH_2CH_2CH_3$	**hept**ane
C_8H_{18}	$CH_3CH_2CH_2CH_2CH_2CH_2CH_2CH_3$	**oct**ane
C_9H_{20}	$CH_3CH_2CH_2CH_2CH_2CH_2CH_2CH_2CH_3$	**non**ane
$C_{10}H_{22}$	$CH_3CH_2CH_2CH_2CH_2CH_2CH_2CH_2CH_2CH_3$	**dec**ane

Cracking. Cracking refers to the process of breaking hydrocarbons into smaller hydrocarbons either by using extreme conditions of temperature and pressure (thermal cracking) or by the use of catalysts with less severe conditions (catalytic cracking). A simple illustration of cracking is the conversion of ethane (C_2H_6) to produce ethylene (C_2H_4) and hydrogen (H_2).

$$\text{H}-\underset{\underset{\text{H}}{|}}{\overset{\overset{\text{H}}{|}}{\text{C}}}-\underset{\underset{\text{H}}{|}}{\overset{\overset{\text{H}}{|}}{\text{C}}}-\text{H} \rightarrow \underset{\underset{\text{H}}{|}}{\overset{\overset{\text{H}}{|}}{\text{C}}}=\underset{\underset{\text{H}}{|}}{\overset{\overset{\text{H}}{|}}{\text{C}}} \;+\; \underset{\underset{\text{H}}{|}}{\overset{\overset{\text{H}}{|}}{\text{ }}}$$

Ethane $\rightarrow$ Ethene + Hydrogen

Ethylene (also called "ethene") is a very important building block in the chemical industry.

The British Thermal Unit (BTU). Sometimes the heat content of the gas is a more relevant measurement tool than measurement based upon volume because the measurement by heat content permits energy sources to be compared on an equal basis. For example, if you wish to use an energy source to heat your home, the amount of heat energy that is provided by the source will be highly relevant (and the volume consumed will be consequential). In some jurisdictions, which exclusively use the metric system, the SI version, the Joule, is used instead of the imperial unit, the BTU.

Product	BTU/Gal
Ethane	65,897
Propane	90,875
Normal Butane	102,950
Isobutane	98,924
Natural Gasoline*	110,020

*Normal Pentane

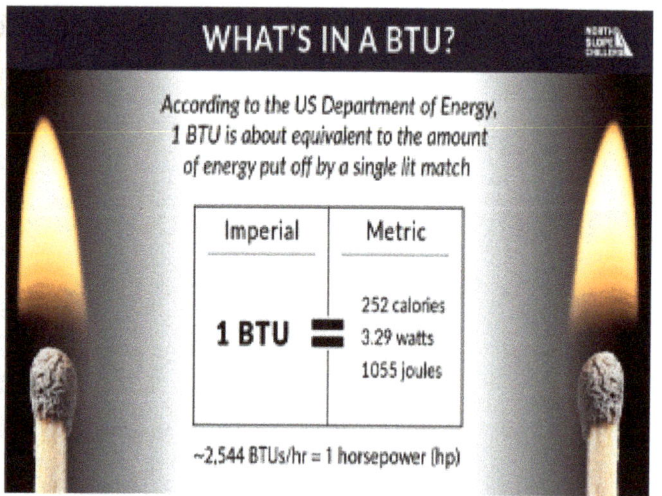

In Trinidad and Tobago, the natural gas that is produced off the North Coast contains methane predominantly, contrasting the composition of the gas that is usually obtained from some fields off the East Coast (which contains many more of the heavier hydrocarbon components). When natural gas is sold on a BTU basis, it allows the seller to benefit from the presence of heavier hydrocarbon components in the gas stream.

Orders of Magnitude. Within the oil and gas industry, volumes are sometimes qualified by letters. For example, T is the equivalent of one trillion, B represents one billion, MM equals one million, and M means one thousand. Any of these letters can appear before certain terms to describe the relevant quantities in cubic feet: MMscf, Bscf, and Tscf. One trillion is written as 1,000,000,000,000 in the international number system, and one billion is written as 1,000,000,000.

Orders of Magnitude		
Hundred	100	2 zeroes
Thousand	1,000	3 zeroes
Million	1,000,000	6 zeroes
Billion	1,000,000,000	9 zeroes
Trillion	1,000,000,000,000	12 zeroes
Quadrillion	1,000,000,000,000,000	15 zeroes

When large quantities of natural gas are under consideration, it is more appropriate to use the larger unit. For example, the volume of the gas in a reservoir would be expressed in terms of Tcf in much the same way that one would refer to the price of an expensive item in dollars ($), not cents (¢).

Temperature. Most people will be aware of the older temperature units, such as degrees Fahrenheit and degrees Celsius, which are still used in the oil and natural gas industries. Degrees Celsius (formerly called centigrade) is an SI-derived unit. The SI temperature unit is actually the Kelvin, and the Celsius scale is related directly to the Kelvin scale. Generally, one measures temperature in Celsius or Fahrenheit (and, if necessary, then converts to Kelvin).

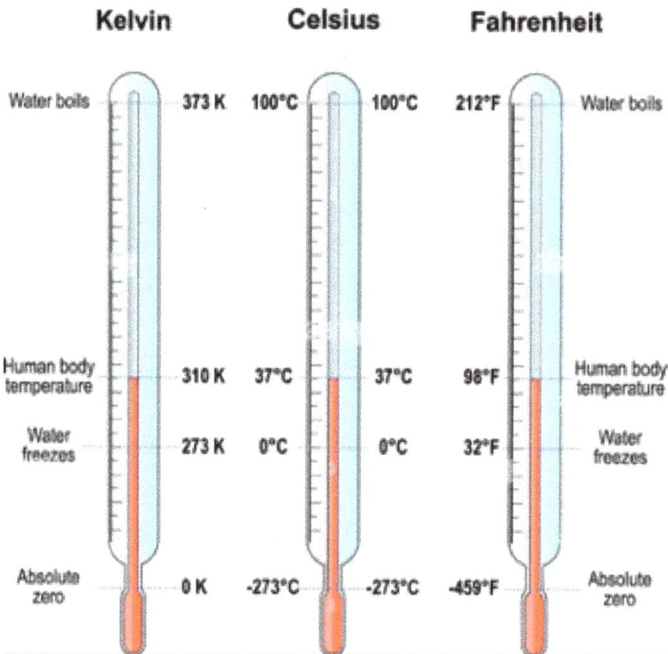

Pressure. The concept of "pressure" in physics is often observed by anyone who has to repair the puncture of a car tyre or introduce air into a soft tyre at a 'gas' station. The old-fashioned unit for pressure is "pounds per square inch" (psi),[2] which is still widely used in the oil and gas industries. The SI unit of pressure is the pascal (Pa). The "bar" is a metric unit of pressure, which is defined as exactly equal to 100,000 Pa (100 kPa).

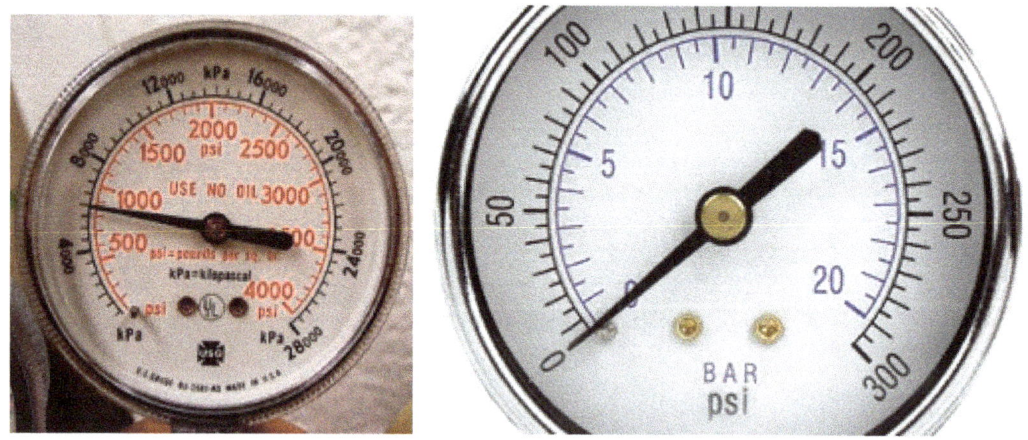

Liquified Natural Gas (LNG). When natural gas (predominantly methane) is cooled to -162°C (-260°F), it becomes a liquid. This liquid, which occupies a significantly smaller volume than the

---

[2] psig (psi "gauge") is a pressure measurement that is relative to ambient atmospheric pressure as distinct from psia (psi "absolute"), the total pressure, which includes 14.7 psi, the value of 1 atmosphere at sea level.

original natural gas, is relatively easy and safe to store in tanks and then transport by ship.

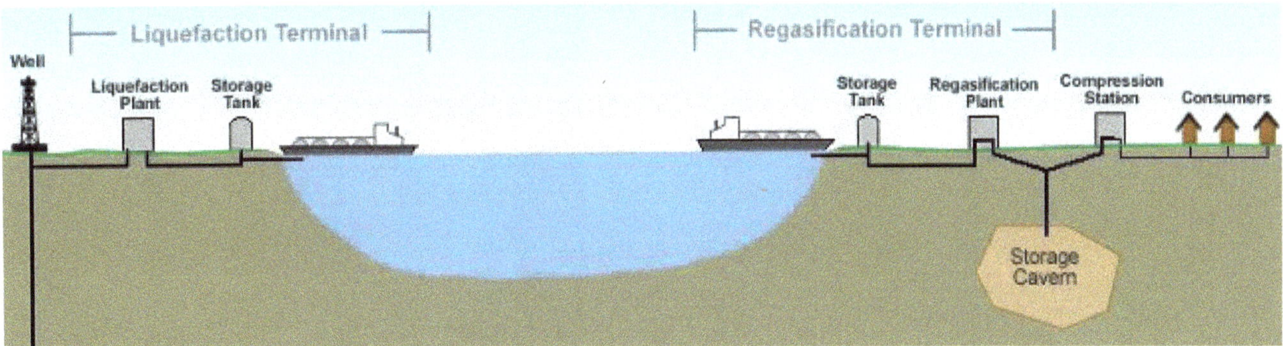

At Point Fortin, LNG is produced, stored in tanks, and later shipped to other parts of the world. At the destination, the LNG is converted to natural gas by a process called "regasification."

Liquified Petroleum Gas (LPG). LPG is a mixture that is composed of mainly propane and butane, which are alkanes that are obtained during the processing of both natural gas and crude oil. Under moderate pressure, these hydrocarbon gases can be liquified and sold in metal cylinders to be used in homes for cooking and also as a reliable energy source for businesses. LPG is also widely used in the aerosol industry, where it is considered an ideal propellant.

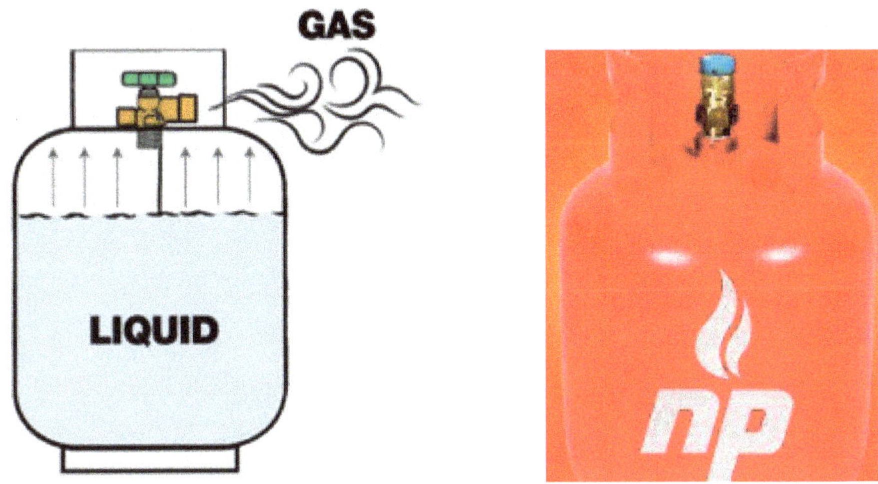

Compressed Natural Gas (CNG). CNG is natural gas that is kept under high pressure. In Trinidad and Tobago, CNG is promoted by NGC and used by consumers as an eco-friendly alternative to conventional fuels for vehicles.

CNG produces engine power when mixed with air and fed into the combustion chamber of an engine. Vehicles can also use natural gas in its liquified form (i.e., LNG), but vehicles in Trinidad and Tobago use CNG that is stored in a cylinder at the back of the vehicle, like the gasoline tank.

Natural Gas Liquids (NGLs). NGLs refer to some of the liquid hydrocarbons that are obtained after natural gas is processed. These hydrocarbons consist of a mixture of ethane, propane, butane, isobutane, and pentane. Thus, LPG (propane and butanes) are also NGLs.

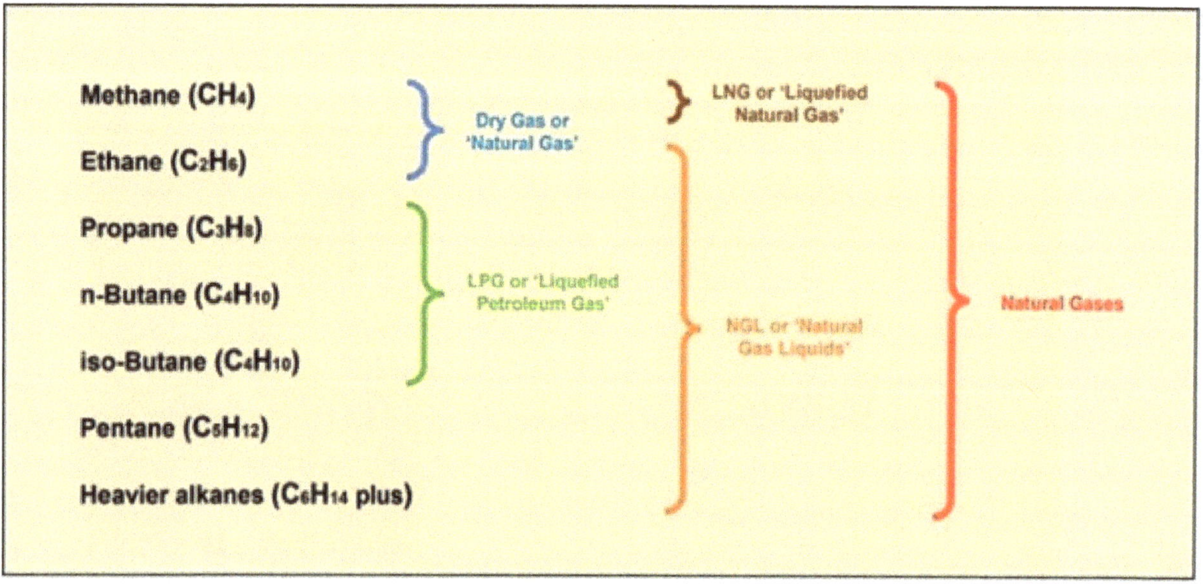

Phoenix Park Gas Processors Limited processes NGLs to produce LPG and heavier liquid hydrocarbons, which are sold internationally and are used as feedstock to produce chemicals.

Petrochemicals. Petrochemicals are chemical products that are ultimately manufactured from crude oil and natural gas. "Base" petrochemicals are chemical intermediates for producing other petrochemicals.

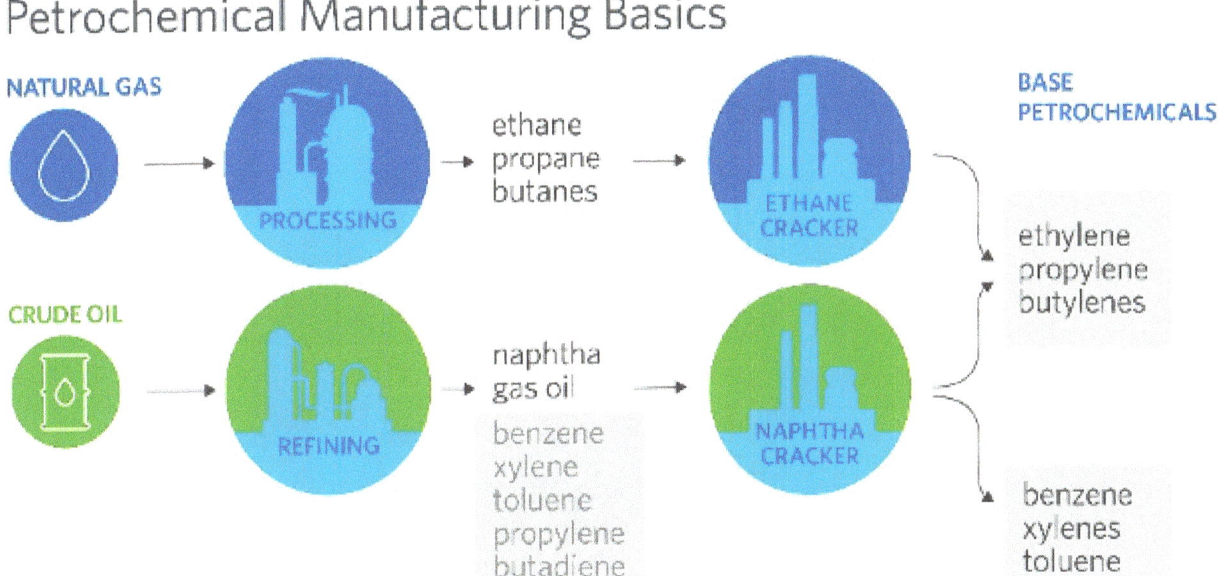

Downstream petrochemicals are produced when a basic petrochemical (for example, methanol) is used as the raw material, a chemical intermediate, to manufacture other petrochemicals.

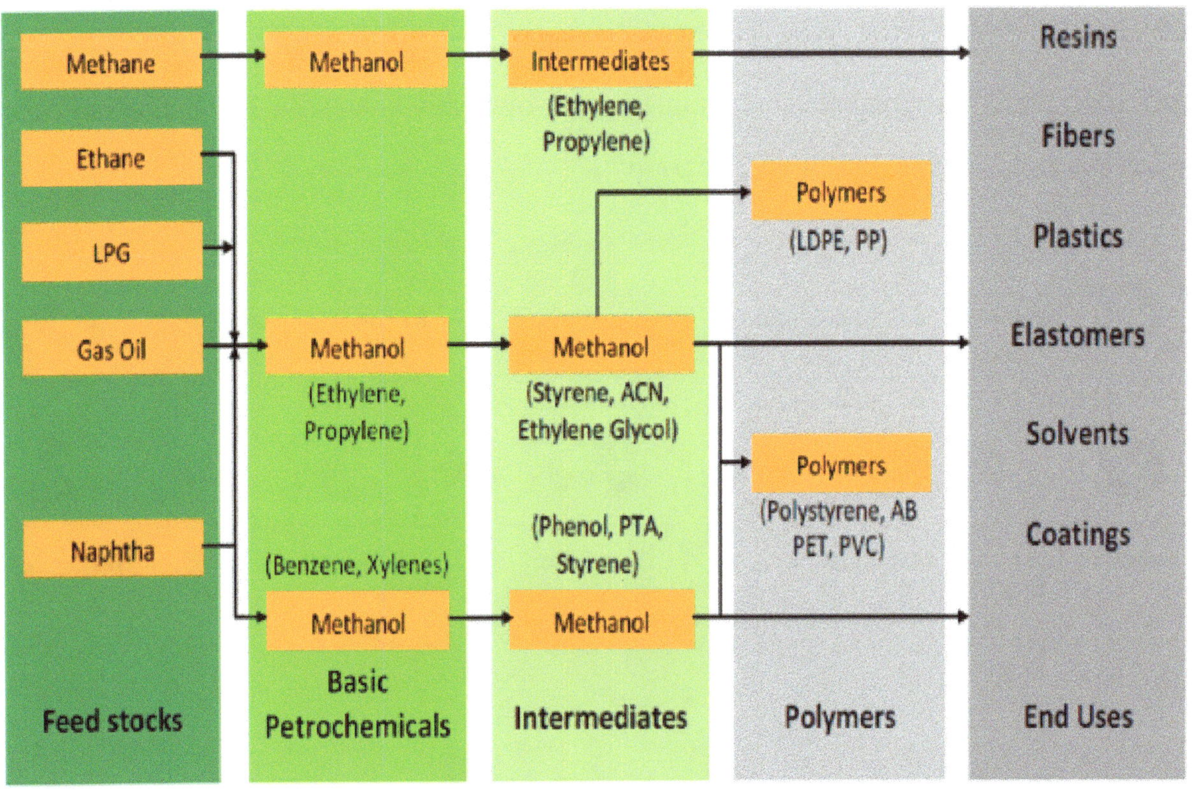

The Value Chain. A value chain is a chain of activities involving the processing of raw materials through a number of intermediate stages or linkages to a marketable end-product. In the natural gas industry, these linkages are typically described as upstream, midstream, and downstream, and they reflect a company's business involvement along the chain.

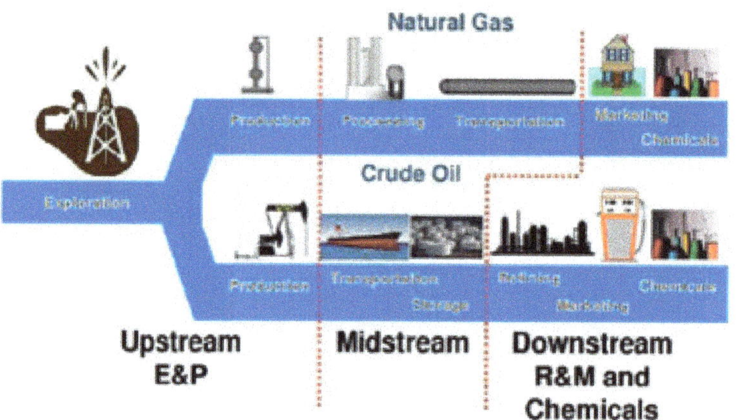

Vertical Integration. A company is said to become more "vertically integrated" when it expands its business (forwards or backwards) along the value chain. If, for example, the producer of a raw material decided to expand its business along the value chain by using its raw material to manufacture products, it would reflect forward integration. On the other hand, if a manufacturer decided to expand the company by producing its own raw materials for the manufacturing operations, it would reflect backward integration.

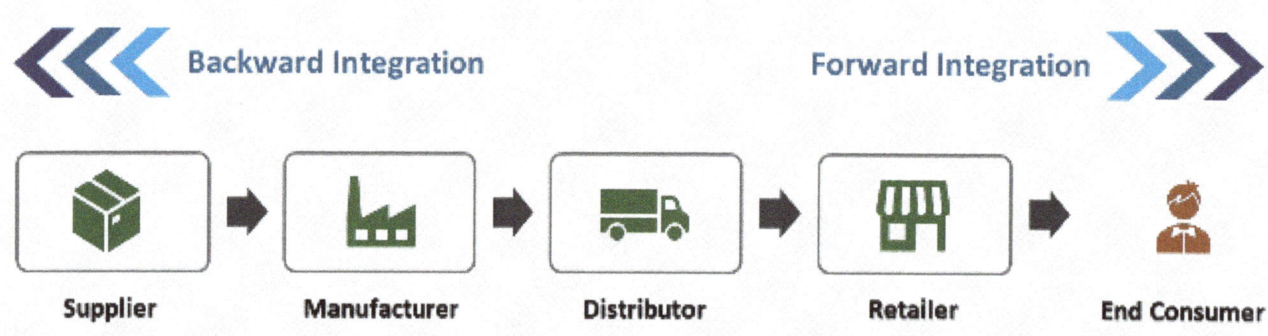

Diversification. The simple rationale is that it is safer to avoid putting all your eggs in one basket.

Pressure, Volume, Temperature ("PVT") Relationships. Water boils at 100°C at normal atmospheric pressure (14.7 psi). In a pressure cooker, which has a constant volume, on heating, the pressure inside the cooker is significantly increased. According to a principle in nature, higher pressures result in higher temperatures when the volume is constant. Cooking is, therefore, faster than at atmospheric pressure with an open pot. The PVT relationships are some of the tools that assist petroleum engineers to assess the behaviour of the hydrocarbons in reservoirs.

Chapter 1: The Energy Sector – A Brief Overview

The history of the oil industry in Trinidad has been well documented by many authors.[1] It is one of the oldest petroleum provinces in the world. There was activity in the exploring and production of the Nation's hydrocarbon treasures since 1857 when the first well drilled for oil, sixty-one (61) metres deep in the vicinity of the Pitch Lake at La Brea, which is located in the southern peninsula of Trinidad. The discovery of oil in La Brea in 1857 created interest among a first wave of companies such as Apex (Trinidad) Oilfields Limited, United British Oilfields of Trinidad ("UBOT"), and Trinidad Leaseholds Limited. In the early 1900s, many refineries were established, including the Point Fortin refinery owned by UBOT in 1912 and in 1917, the Texaco Refinery at Pointe-à-Pierre. Trinidad was even perceived as providing security of supply for Britain. As one key pioneer observed:

"It has also been manifest that Great Britain, before adopting oil as fuel for her navy, must be sure of a source of supply which the outbreak of war will not cut off, and that it is therefore

---

[1] The information in this chapter is based mainly upon the data provided by Trevor Boopsingh in his book entitled, 'From Walter Darwent to Atlantic LNG Train 4', https://books.google.co.uk/books?id=W9k-BAAAQBAJ&pg=PA1&source=gbs_toc_r&cad=3#v=onepage&q&f=false and the Eric Williams Memorial Lecture by given by Professor Ken Julien in 2005 https://www.central-bank.org.tt/sites/default/files/lectures/19th%20Dr.%20Eric%20Williams%20Memorial%20Lecture.pdf.

important that sources of supply should, if possible, be found on British territory… In the light
of my discoveries, I felt that Trinidad, England's most valuable possession in the West Indies,
being as it is one of the keys to the Panama Canal, now rapidly approaching completion, might
herself one day be one of the chief sources of supply of oil fuel, and thanks to that and her
unique position, might become one of our most important naval bases … ."

By 1926, Trinidad and Tobago had become the largest oil producer in the British Empire, and during World War II, the Nation's hydrocarbon resources came firmly under the control of the U.S. and the U.K. As early as 1955, Dr Eric Williams, Trinidad and Tobago's 1st Prime Minister, observed that, with respect to oil,

".. all the evidence, from the earliest times, has indicated a subordination of local interests
to those of external capital." Dr Williams also recognised that *"….. there will come*
occasions when the State may have to take the initiative as an investor, without prejudice
to the policy of encouraging and supporting private enterprise, in order to protect and
promote the National interest."

A joint venture with Tesoro Corporation to acquire the local producing assets of BP was the first such bold step of State ownership in strategic industries. And later, on 31st August 1974, the oil industry was locally owned when the Government purchased Shell's assets to create the Trinidad and Tobago Oil Company Limited (TRINTOC), the country's national oil company. Dr Williams, referring to the acquisition of Shell's assets, illustrated the strategic importance of that acquisition:

"The real question was not whether the oil flowed ….. but to whom did the benefits flow –
U.B.O.T. ….. United British first, Trinidad last. Then the name was changed …… Shell
first, Trinidad last …. Reverse that, put Trinidad and Tobago first – TRINTOC. Now we
know not only from where the oil flows, but to whom the benefits flow."

The oil industry was later beset by rising costs, competition from supplies of low-cost crude from the Middle East and Africa, and the construction of new refineries in Europe and the United

Kingdom.[2] Consequently, a decline in oil production and oil prices led to withdrawal from the domestic energy sector by Shell, BP, and Texaco and ultimately culminated in the formation of the State-owned Petrotrin.

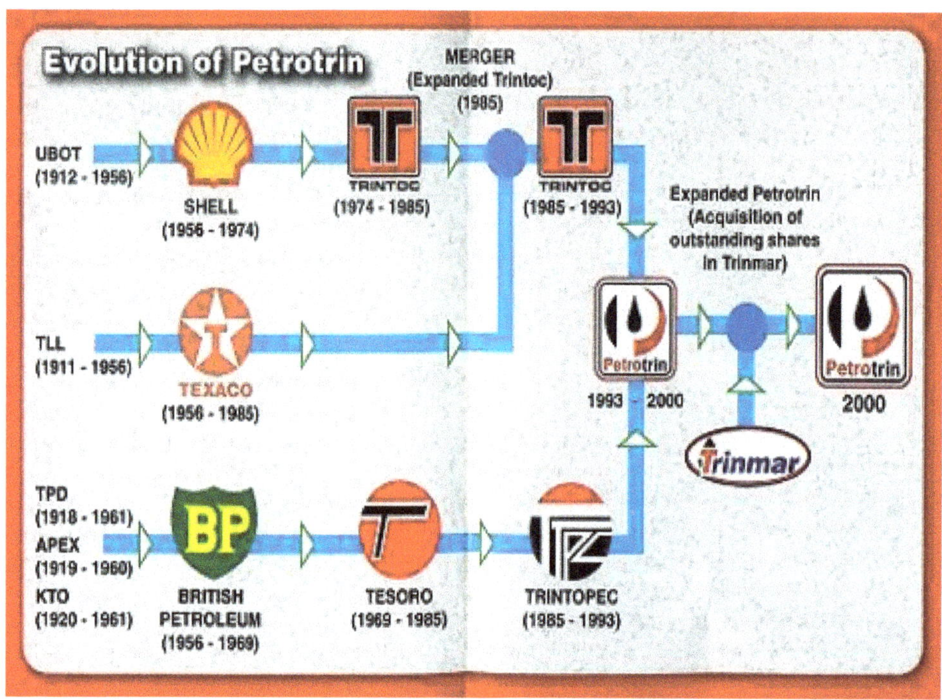

The Vision of the Energy Sector. On 17th January 1976, when the Government decided to invest in the Iron and Steel Company of Trinidad and Tobago ("ISCOTT"), some of Dr Williams' words said more about the vision for the National Energy Sector than about ISCOTT:[3]

> " .. the basis of the policy for colonial development, the classic exposition of which was the prohibition on the colonies, especially the mainland colonies in America, expressed by a British Prime Minister, [was] that the colonies were to manufacture not a nail, not a horseshoe. They were to produce raw materials only, which were to be sent to England, to enable downstream manufacturing operations, to provide jobs, to expand."

Dr Williams continued:

---

[2] Trevor Boopsingh, 'From Walter Darwent to Atlantic LNG Train 4', *supra*.
[3] Per Ken Julien's Defining Moment No. 7, The 19th Memorial Eric Williams Lecture, at p. 20.

"There have been attempts to persuade us that the simplest and easiest thing to do would be to sit back, export our oil, export our gas, do nothing else and just receive the revenues derived from such exports and, as it were, lead a life of luxury – at least for some limited period. This, the Government has completely rejected, for it amounts to putting the entire nation on the dole. Instead, we have taken what may be the more difficult road, and that is, accepting the challenge of entering the world of steel, Aluminium, methanol, fertilizer, and petrochemicals. We have accepted the challenge of using our hydrocarbon resources in a very definite industrialization process."

The NEC. The National Energy Corporation of Trinidad and Tobago ("NEC") grew out of a Coordinating Task Force (CTF), which was set up to consider how to monetise the country's natural gas resources. The NEC was incorporated in 1979 to continue the work first started by the CTF, and in 1980, it described its role in the following terms:

"The role of the National Energy Corporation is to guide the development and management of oil, gas, and other mineral resources of Trinidad and Tobago; to assist the government in the formulation of energy and industrial policy and strategy; and to increase public participation in decision-making in key areas of the resources and entry-related industries."

Pipeline Infrastructure. At the embryonic stages of the transformation to natural gas in the energy sector, there were no natural gas offshore or cross-country pipelines, which meant that the challenges to monetise the Nation's natural gas reserves were, as described by Dr. Julien, "forbidding and formidable." The next phase was, therefore, the development of the appropriate pipeline infrastructure to bring gas to Point Lisas. The National Gas Company of Trinidad and Tobago ("NGC") was thereby established in 1975.

In 1980, the Government used its option in the exploration and production licences under which Amoco had operated, to collect all flared gas without having to pay for it. NGC subsequently built two compressor platforms next to Amoco's offshore producing installations in the Teak and Poui fields, where over 180 MMscf of natural gas was being flared and wasted daily.

Thus, initially, NGC had a fairly narrow remit of gas transmission, distribution, and compression, whereas the NEC functioned as the implementation arm for the Government's strategy, with a focus on business and infrastructural development work.

Transformation. Following the discoveries of large reservoirs of natural gas, priority was given to industrial diversification fueled by natural gas. Through a combination of State equity and foreign direct investment, Trinidad and Tobago witnessed the expansion of the ammonia industry and the birth of the steel and methanol industries. The first natural gas customer was Tringen, a joint venture between W.R. Grace and the Government. Subsequently, in 1981, Fertilizers of Trinidad and Tobago Limited (Fertrin), a joint venture between the Government (51%) and Amoco (49%), commissioned two ammonia plants, each with a rated capacity of 344,500 tonnes per annum.

The new emphasis on gas was demonstrated in 1982 when the Government deferred the refinery upgrade of the former Shell refinery at Point Fortin and opted to approve the construction and financing of fully State-owned methanol and urea plants in 1984. The Trinidad and Tobago Urea Company plant, was commissioned in 1984. Urea production maximised the value-added from the investment in ammonia. The first methanol plant was also commissioned in 1984 and began commercial production in 1984. In April 1994, after the economic recession of the 1980s, the Point Fortin refinery was shut down, which later made way for Atlantic LNG, thereby reflecting Trinidad and Tobago's continuing shift in its emphasis from oil to natural gas during the 1990s.

The Private Sector. A new Government took office at the end of 1986, and it identified (among other things) divestment as a tool for structural transformation to address the Government's need for financial resources. The State-owned methanol and urea plants were sold to the private sector, and one of the policies of this new Government was the use of a more open approach to private capital. The State's involvement in natural gas-based petrochemicals declined after the early 1990s, and Trinidad and Tobago subsequently became a major natural gas development centre in the world.

The NGC/NEC Merger. In the early 1990s, NGC merged with the NEC, and NGC assumed a widened mandate to promote further natural gas-based development. The vision for the energy sector, which was then being implemented by the NEC, was to increase the valued-added of the 1st stage petrochemicals by the creation of further downstream natural gas-based petrochemicals. The rationale was that, over time, the local availability of those downstream petrochemicals would have facilitated greater industrialisation by deepening the petrochemical sector.

Petrochemical Growth. The ammonia production grew to eleven ammonia plants that included two ammonia complexes with a total annual capacity of 5.2 million metric tonnes (MT). Since

1984, the methanol industry has also expanded to include six larger plants with an annual production capability of over 6 million MT of methanol. **One of the companies, Methanol Holdings (Trinidad) Limited (MHTL),** part of the Proman family of companies, became one of the largest methanol producers in the world with a total capacity of over 4 million MT annually from its five methanol plants. The company was the largest supplier of methanol to North America and was also a significant supplier to the European Market.

LNG. Over time, the production of petrochemicals for export by the private sector, without the manufacture of local valued-added downstream derivatives, represented just another method of exporting the natural gas raw material. Accordingly, there was an increasing policy preference to utilise gas for LNG where there were policy benefits arising from larger upstream gas production:

(a) The availability of large volumes of ethane would potentially have facilitated investment in an ethylene complex with significant downstream value-added opportunities.
(b) There would be an increase in Government revenue from upstream taxation.
(c) The NCMA gas reserves were stranded, and it was necessary to produce large volumes of gas to develop these reserves and the associated pipeline infrastructure economically.

In 1999, with the introduction of the Atlantic LNG Company of Trinidad and Tobago, a joint venture among, Boston-based Cabot LNG, NGC, Amoco, British Gas, and Repsol, Trinidad and Tobago became an LNG-producing country. And, in December 2005, after the construction of the 4th LNG Train, which was, at that time, the largest in the world, the Atlantic LNG plants reflected one of the largest global liquefaction facilities. The Caribbean Nation also became the largest supplier of LNG to the United States and currently supplies LNG to various global destinations, including Spain. Currently, the two largest stakeholders in these four LNG Trains are Shell and bp, who both supply gas for the production of LNG at Point Fortin and for petrochemicals at Point Lisas.

Gas Reserves. The natural gas reserves have been in decline over the last decade as the rate of reserve additions has failed to keep pace with consumption.

Chapter 2: My Introduction to the Energy Sector.

I entered the energy sector in 1980 when I joined Trintoc. I was recruited to the NEC in 1984. Trintoc had upstream operations at Penal and midstream refinery operations in Point Fortin. I was very involved with the chemistry of crude oil processing with Richard Callender, the laboratory chemist at the Point Fortin refinery; later, at Penal, with the principles that govern the properties of petroleum reservoir rocks and fluids. From a natural gas perspective, these experiences were critical precursors for my insights into project planning and development recommendations for natural gas projects while I was at the NEC. The Trintoc refinery site at Point Fortin became the location for the manufacturing operations of Atlantic LNG after the refinery was shut down.

Fractional Distillation. The Point Fortin refinery was very small compared to the Texaco refinery, which I had visited as a schoolboy. In Point Fortin, the refinery products were liquefied petroleum gas ("LPG"), gasoline, naphtha, kerosene, diesel oil, lubricating oil, fuel oil, and bitumen. The bitumen was used locally, in conjunction with pitch from La Brea, for road surfacing, and it was considered to be excellent for that purpose because of its predictable consistency.

"

As a research chemist, I was very interested in the components of crude oil because some of these components (for example, aromatics) are important feedstocks for the production of intermediates in the chemical industry.

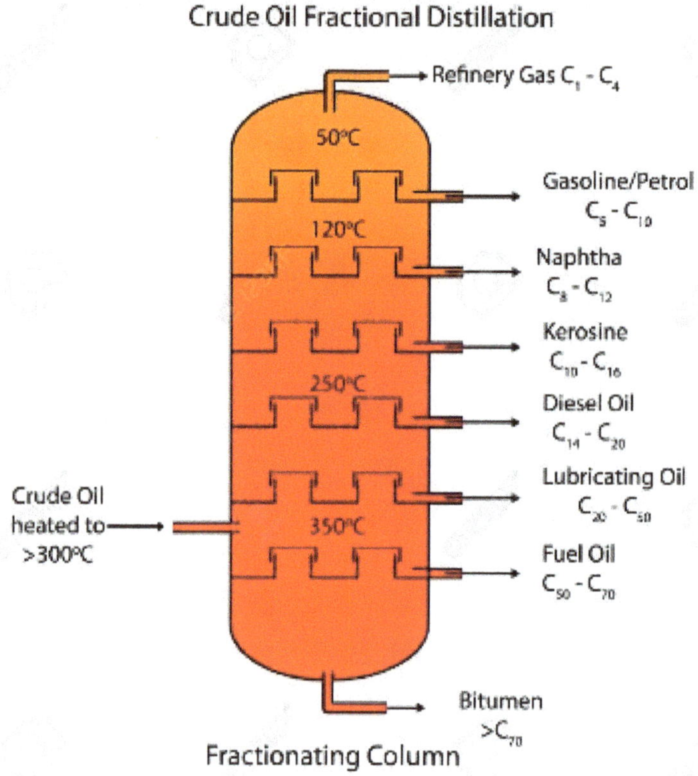

Crude Oil Fractional Distillation

Petroleum Fractions

Some of the petroleum fractions produced from distillation columns with their boiling point ranges are given in Table below. These fractions may go through further processes to produce desired products.

Petroleum fraction	Approximate hydrocarbon range	Approximate boiling range	
		°C	°F
Light gases	C_2–C_4	−90 to 1	−130–30
Gasoline (light and heavy)	C_4–C_{10}	−1–200	30–390
Naphthas (light and heavy)	C_4–C_{13}	−1–205	30–400
Jet fuel	C_9–C_{16}	150–255	300–490
Kerosene	C_{11}–C_{14}	205–255	400–490
Diesel fuel	C_{11}–C_{16}	205–290	400–550
Light gas oil	C_{14}–C_{18}	255–315	490–600
Heavy gas oil	C_{18}–C_{28}	315–425	600–800
Wax	C_{18}–C_{36}	315–500	600–930
Lubricating oil	>C_{25}	>400	>750
Vacuum gas oil	C_{28}–C_{55}	425–600	800–1100
Residuum	>C_{55}	>600	>1100

Aromatics. Aromatics are cyclic compounds that are produced by the removal of hydrogen from the naphthenes in the naphtha distillation fraction. The removal of hydrogen is a process called "dehydrogenation."[4] Cyclohexane, for example, is dehydrogenated to form benzene.

$$\text{Cyclohexane} \xrightarrow[\Delta]{\text{Catalyst}} \text{Benzene} + 3H_2$$

A mixture of cyclohexane and its derivatives can be converted into a mixture of aromatics such as benzene, toluene, and xylenes (often referred to as "BTX"). This mixture of aromatics can be separated by fractional distillation into individual components. The Point Fortin refinery (unlike the refinery at Point – à- Pierre) did not have a BTX separation unit.

benzene   toluene

ortho-xylene   meta-xylene   para-xylene

Aromatics are important refinery products because they are building blocks in the chemical industry for many chemicals and polymers, such as phenol, tri-nitrotoluene, nylons, and plastics.

---

[4] The reverse process, i.e., the addition of hydrogen, is called "hydrogenation". At the Point Fortin refinery, in order to meet the smoke point specification for jet fuel (which is kerosene), the kerosene stream was hydrogenated. The aromatics in that stream were thereby converted into naphthenes. The smoke point is an important product quality specification for jet fuel and other grades of kerosene because it is a measure of a fuel's tendency to generate smoke when burned. Smoke is caused by the formation of carbonaceous particles that do not completely combust, which are concerns, both because of their effect on the environment, and on the engine.

API Gravity and Specific Gravity. Several methods have been utilised to differentiate the characteristics or properties of crude oils, including the American Petroleum Institute (API) gravity number and the specific gravity of the crude oil.

Crude Oil Type	API Gravity
Light	Larger than 31.1°
Medium	Between 22.3° and 31.1°
Heavy	Smaller than 22.3°

One of Trintoc's heaviest crude oils was the "Guapo" crude, which, with an API gravity in the vicinity of 15°API, was very viscous, almost like molasses. The Vessigny/Guapo/Parrylands fields in the Point Fortin area are believed to contain substantial quantities of heavy oil.

Benchmarks. 'Sweet' crude oils with high API gravities generally fetch a premium[5] because they are highly desirable for refining into high-quality gasoline, diesel, and other fuels. The source of crude oil is also important because lower transportation costs to deliver the product ultimately make it cheaper for the consumer. Because of these factors, buyers of crude oil use "benchmarks" that provide an easy way to value the commodity based on its quality and location. The Brent crude and North America's West Texas Intermediate crude (WTI) are benchmarks.

Penal. After a year in Point Fortin, as a consequence of the company's shifted emphasis to subsurface research, I began to work in Penal within the petroleum engineering group. Even then, the late Hon. Senator Franklyn ("Frankie") Khan, former Minister of Energy and Energy Industries, one of the geologists, had his characteristic 'smile'.

In 1982, with technical assistance from Braspetro, Trintoc discovered the Chaconia gas field in the North Coast Marine Area ("NCMA") off the North Coast of Trinidad.

---

[5] But some heavy sweet crudes with low API gravities can be attractive because of their blending utility in reducing the pour point.

I accompanied two reservoir engineers, Alan Russell and Dave Mohammed, on the historic well test of this field where the gas from a well, located at 500 ft. water depth, was flowed for 36 hours without a decrease in the initial reservoir pressure.

In Penal, my main function was to develop a research programme based upon the physical properties of the various reservoir fluids, such as formation water and hydrocarbons produced from indigenous petroleum fields. Thus, I became intimately familiar with the quality of the various crude oils produced from Trintoc's land operations where, for example, the heavy crudes in Parrylands contrasted the light wax-producing crudes from the Catshill and Balata fields.

By agreement, as part of Trintoc's commitment to the industry, I taught, part-time, a course on the "Properties of petroleum reservoir rocks and fluids" in the Petroleum Engineering Unit at UWI, St Augustine. Teaching the scientific principles (some of which are illustrated below) associated with this subterranean science paled into insignificance when compared with the very difficult task of reservoir engineers. They had to develop mathematical models to predict the behaviour of the hydrocarbons in the petroleum reservoirs and quantify the volume of the oil and gas in place.

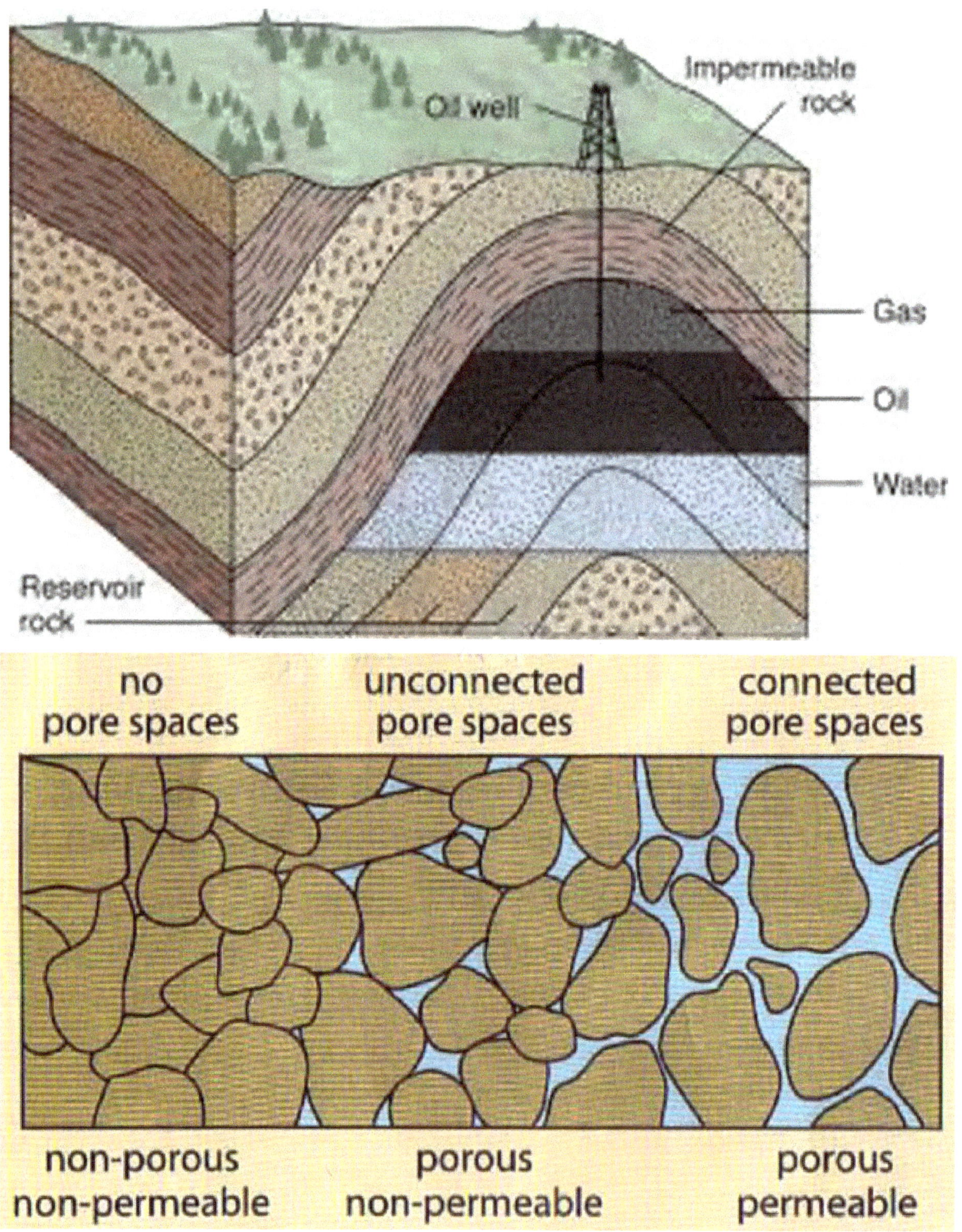

Wettability. The tendency of a liquid to spread over the surface of a solid provides an indication of the *wetting* characteristics of the liquid for the solid.

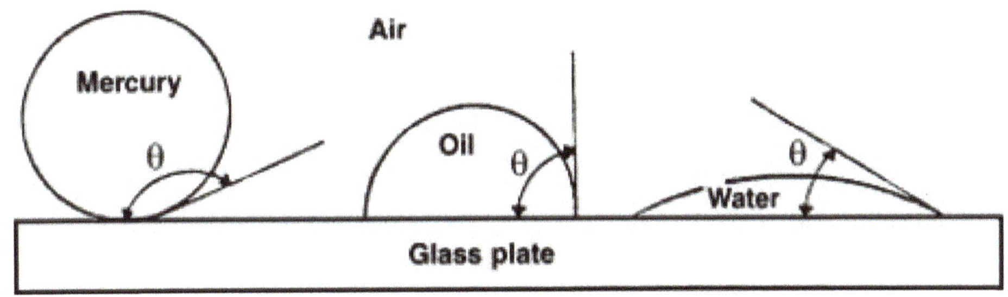

In water-wet conditions, a thin film of water coats the surface of the rock, which is a desirable condition for efficient oil transport. Accordingly, chemical treatments that change the wettability of the rock formation from water-wet to oil-wet can lead to the retention of oil in the reservoir (significantly impairing productivity). In an oil-water system, oil is normally the non-wetting phase, whereas, in the presence of gas, the oil is the wetting phase. The minute film of the wetting phase on the rock surface does not contribute any flow capacity for the wetting phase but decreases the volume that is available for the non-wetting fluid.

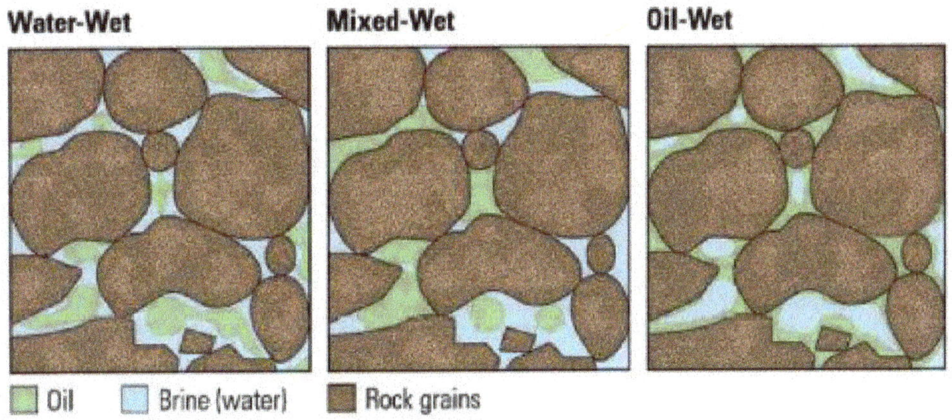

PVT Relationships. In the laboratory, when a gas is increasingly compressed at a fixed temperature, liquid begins to condense at a point called the "dew point." The dew point represents a transition from the gaseous state to the liquid state. In the 2-component system below, at the dew point, the system will consist of two phases: vapour and an infinitesimally small amount of liquid. As the pressure is further increased and the system continues to contract isothermally, more and more liquid will condense until the system reaches a point called the "bubble point." At the bubble point, the system will consist almost entirely of liquid except for an infinitesimal amount of vapour. PVT relationships can help to estimate the phase behaviour of multicomponent systems such as the complex hydrocarbon mixtures found in petroleum reservoirs.

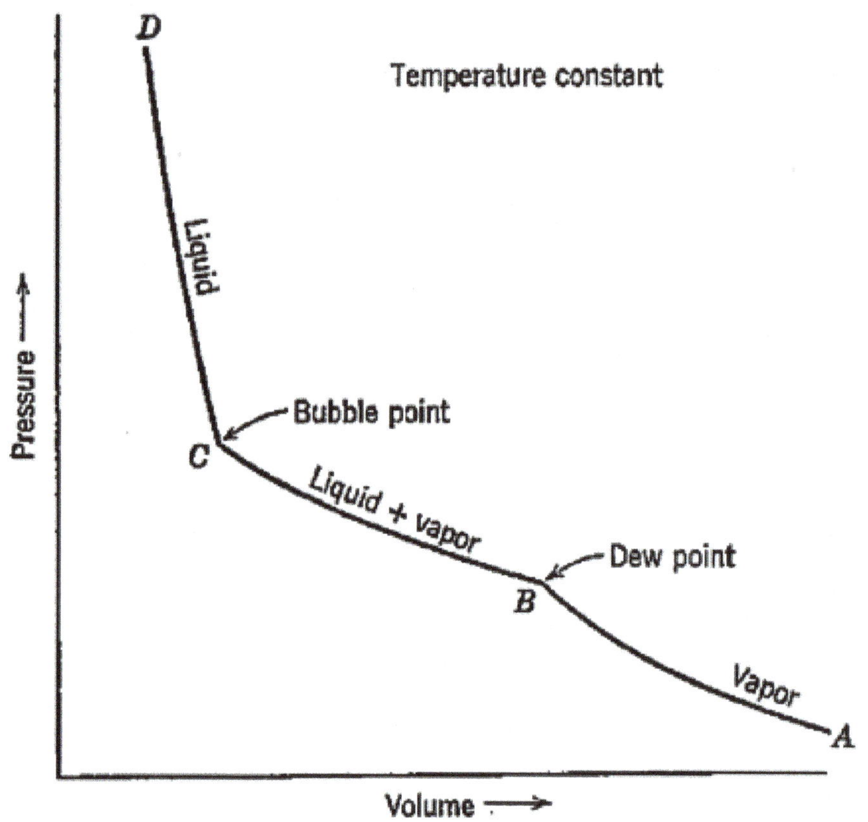

Critical Temperature. A characteristic of gases is that there is a temperature above which its vapour cannot be liquefied, regardless of the applied pressure. This temperature is called the "critical temperature." Point C in the diagram below represents the critical temperature.

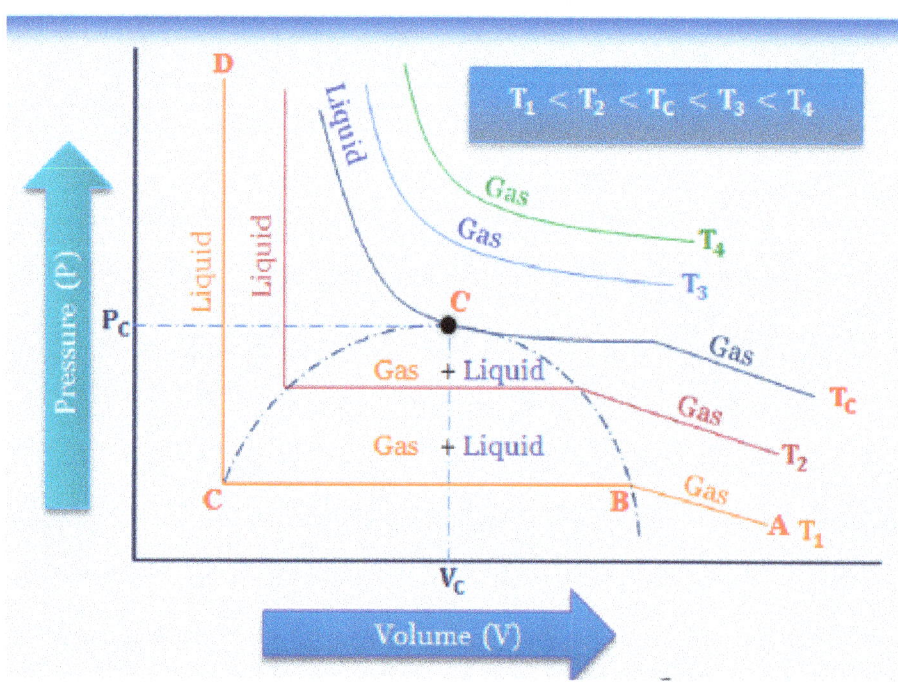

Gases become more difficult to liquefy as the temperature increases because the kinetic energies of the particles that make up a gas also increase. At (and above) the critical temperature, the energy of the particles keeps them sufficiently apart so that they can no longer be liquified. In the previous diagram, along the isotherms T_3 and T_4, which are greater than the critical temperature (Tc), only the gaseous phase exists because vapour does not attain its dew point above the critical temperature. The critical temperature of methane is *circa* - 82°C, which means that a petroleum reservoir consisting predominantly of methane, normally at high temperatures, would always exist in the gaseous phase regardless of the reservoir pressure. Contrast propane and butane, which have critical temperatures of 96.8°C and 152°C, respectively. These hydrocarbons can be liquids at ambient conditions.

Gas Production. Natural gas can occur by itself as a single phase in a reservoir or in conjunction with liquid petroleum oils, such as in a "gas cap" where a gaseous phase coexists with a liquid hydrocarbon phase. When this occurs, the composition of the gaseous phase will be very complex at the relatively high temperatures and pressures in petroleum reservoirs.

Temperature and pressure are variables that affect the solubility of substances. And since petroleum reservoirs usually exist at conditions of high temperature and pressure, at the equilibrium conditions in the reservoir, gas will dissolve in the liquid phase, and high-boiling hydrocarbon constituents will dissolve in the gaseous phase. Thus, the gaseous phase in a reservoir with a gas cap will contain several heavy hydrocarbons, which would occur as high boiling point liquids if they were to be isolated as pure substances.

When a high-pressure gaseous phase is saturated with dissolved liquid hydrocarbons, condensate will be produced when the pressure is reduced. This phenomenon is illustrated in the diagram below for the gaseous phase.[6] Points B and Á exist on isotherms representing approximately 15% and 25% liquid compositions, respectively. Thus, as the pressures decrease along the isothermal path AB (or along AÁ towards lower surface temperatures), liquid is produced.

---

[6] D.L. Katz and B. Williams: Reservoir Fluids and their Behaviour, *Bull*. AAPG, 36:342 (1952)

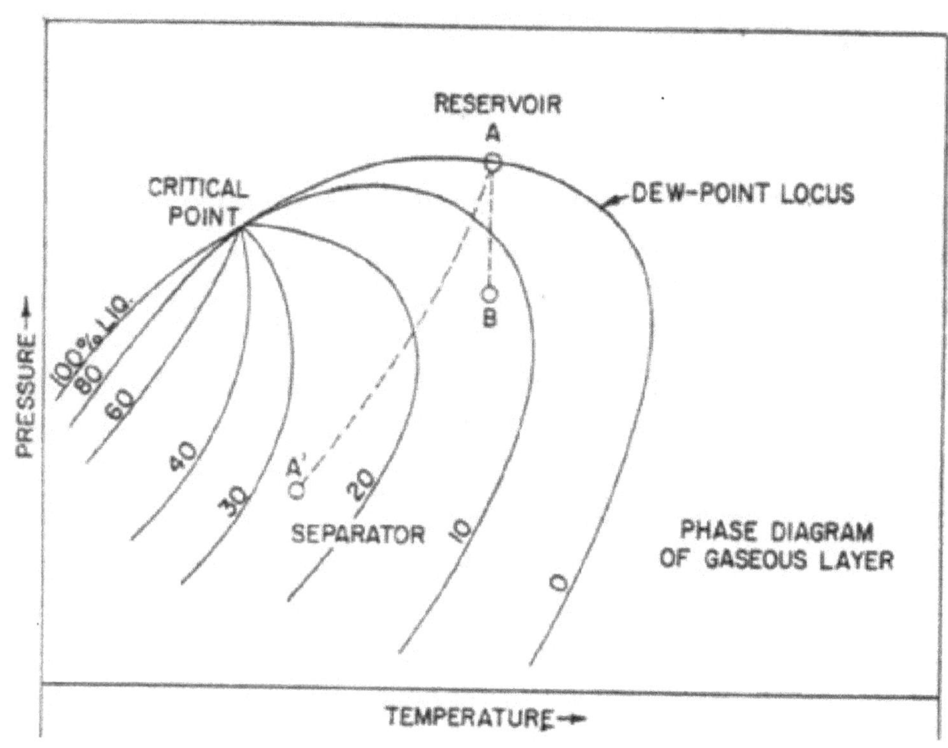

For the liquid phase, gas ("associated gas") is produced when the pressure is reduced either isothermally (path AB) or along path AÁ as the oil moves towards the lower surface temperatures.

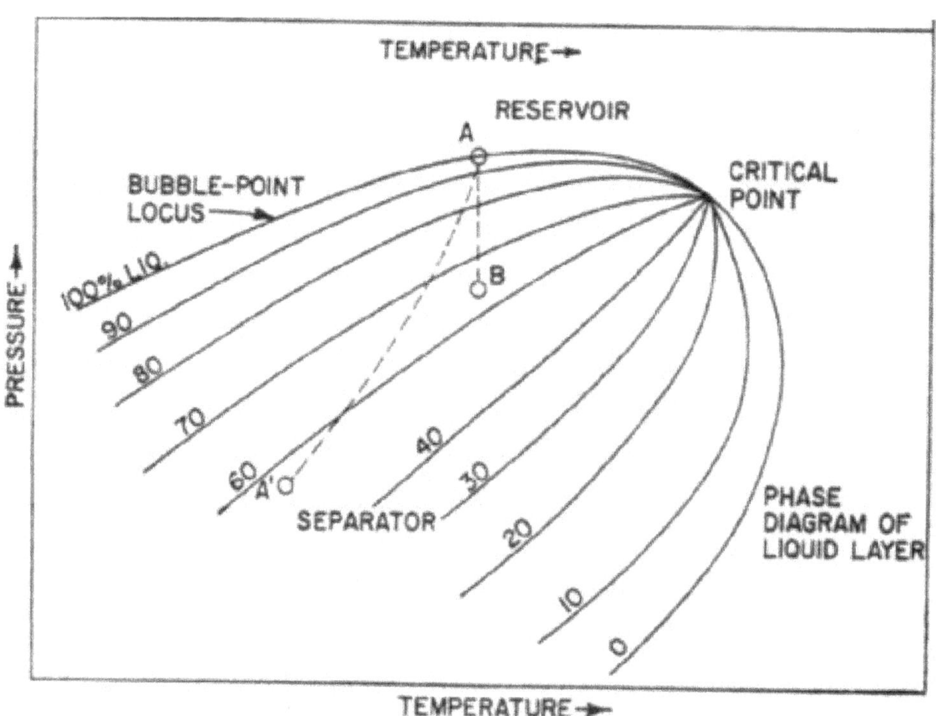

As a result of the previously discussed principles, when natural gas, which is saturated with heavy hydrocarbons (i.e., "wet" gas) flows in pipelines, as the pressure decreases during transportation, the heavier hydrocarbons (i.e. condensate) will increasingly drop out from the gaseous phase.

Reserves Estimates. The foregoing discussion, a simplistic overview, is meant to illustrate the geological and petroleum engineering challenges when quantifying the hydrocarbon resources in subsurface petroleum reservoirs. This uncertainty underscores the rationale for categorising hydrocarbon reserves as proved, probable, and possible, based upon the respective odds of commercial extraction using any combination of the regulatory, economic, and technological challenges to extract these resources profitably.

The NEC. In 1984, an NEC committee, chaired by Dr. Kenneth Julien and including Eldon Warner, CEO, and Basharat Ali, Head Technical Services, recruited me to the NEC from Trintoc.

I joined the NEC as a Senior Project Analyst in the Technical Services Division with effect from 15th August 1984. My office was located at the methanol plant at Point Lisas. Prior to taking up my duties, I began a six-week familiarisation tour with various organisations, including the IDC, the Ministry of Energy, and NGC.

The Ministry of Energy. At the Ministry, where Trevor Boopsingh was then the Permanent Secretary, I was provided with a general overview of the natural gas resources and gas utilisation:

(a) There was 'dry' gas in the undeveloped gas reserves existing off the North Coast, 'wet' gas in the gas-condensate reservoirs off the East Coast, and associated gas from oil production.

(b) The dry gas off the North Coast was high in methane content (about 99.4%) whereas the methane content of the gas in the East Coast fields was about 92.4%.

(c) A Ryder-Scott study had indicated that proved reserves were estimated at 15.55 Tcf and the total reserves (proved, probable and possible) were about 18 Tcf.

(d) Recoverable reserves from the Cassia and Teak fields had been recently estimated at 1601 Bcf of which 728 Bcf had already been produced.

(e) It was contemplated that further development would be required by drilling additional wells in the Cassia 25 and 23 sands if current demand had stabilised at 400 MMscfd.

(f) In 1983, 20% of natural gas production had resulted from land-based companies and TRINMAR. Amoco, whose production was about 480 MMscfd from the Cassia and Teak fields, had accounted for the rest.

(g) Associated gas, compressed by NGC at the Teak and Poui fields, had averaged 57 MMscfd in 1983 and had risen to 83 MMscfd, so far in 1984.

(h) Prior to 1983, apart from the establishment of 1 ammonia-producing facility in 1959, there had not been any significant gas-based petrochemical industries in Trinidad and Tobago.

(i) The Government had provided about U.S.$800 million (reflecting more than two-thirds of the total cost for the projects on stream) for market development and much of the infrastructure for the natural gas industry.

(j) In a continuing trend, gas production, over 81% of which was currently being utilised, had risen from 250 MMscfd in 1973 to 600 MMscfd in 1983.

NGC. At NGC, where Malcolm Jones was General Manager, I was informed that –

(a) Although the major assets (the offshore platforms and major pipelines) were owned by the Government, a decision had been taken to vest these assets in NGC; and

(b) The pipeline infrastructure could have transported 1000 MMscfd.

Gas Sales. The flare gas project had been very successful. Amoco paid NGC 20 cents per Mscf for 30-32 MMscfd of associated gas, which NGC delivered to Amoco at a pressure of 850 psi. This associated gas had yielded more oil for Amoco in its gas-lift operations than if Amoco had used its own high-pressure gas in these operations. In the general gas contract with Amoco, NGC purchased gas at 60 cents per Mscf. That price had been determined by adding 'pass through' increases such as royalties and taxes to a 1983 base price when Cassia came on stream. The maximum contracted quantity was 200 MMscfd.

Liquid Drop out in the Pipeline. While at NGC, I evaluated a problem of liquid drop out in the transmission and distribution systems, and on 17th October 1984, shortly after my return to the NEC, I shared my insights with Malcolm Jones:

"Dear Malcolm

Please accept my sincere appreciation for the most enjoyable and informative week with NGC, which culminated in our stimulating discussion on Friday, 12th October 1984. ... ...

The problem of liquid dropping out in the transmission and distribution systems, particularly at FERTRIN, has encouraged me to make the following preliminary assessment using a gas analysis from CARIRI and a conservative daily flow rate of 350 MM scf. The estimates would, of course, be higher if the supply of gas were increased. Calculations put the daily LPG potential at about 286,206 USG or 395,443.4 KL/year, which could be a strategic energy source if the supply from Point-a-Pierre were either unavailable or unable to adjust to the growing demand. Energy balances for the country suggest that the latter is a real possibility if the current trend continues.

Perhaps an even greater attraction may be found in capturing the naphtha stream, which at 61,300 USG/day could earn about 16.1 million U.S. dollars per annum, or a little more, when further quantities of condensate are brought on stream from the 5th compressor at NGC Poui. Stripping the propane and heavier fractions from the natural gas would decrease the net BTU/scf, but the change (about 7%) would result in a reduced income from T&TEC, your largest customer, of only about 0.5-0.6 million dollars per annum.

The pressure drop and commensurate loss of some of the heavier material during transport across the country, although unattractive, may be an economic necessity since berthing facilities and tank farm storage at Texaco should be available.

The yields for the various fractions would depend upon the process used, and the NEC, with their vast experience in studying liquefaction, may be suitably poised to develop the optimum process for our indigenous fuels.

The Galeota Point Incident. Out of the blue, in December 1984, the NGC onshore pipeline system was flooded all the way to Point Lisas with condensate ("the incident"). Consequently, all the petrochemical plants at Point Lisas were shut down. I was appointed the chairman of a committee comprising NGC, NEC, FERTRIN, the Ministry, and Amoco, to investigate the incident. The committee quickly established, from operational charts of the slug catcher at Galeota Point, that the incident had been caused by an operational failure of the slug catcher. At a committee meeting, NGC reported that four of the five drip pots along the 24′ line between Beachfield and Picton were out of service, and the 5th, which was located in the vicinity of Picton, was filled with condensate every day.

Historically, the gas quality delivered to customers at Point Lisas was a major concern because at the first stage in petrochemical manufacture, the reformer catalysts, for optimum efficiency, were designed to convert the simple alkanes (for example, methane and ethane) to "synthesis gas" - carbon monoxide (CO) and hydrogen (H_2). However, the presence of heavier alkanes (for example, pentane, hexane, etc.) in the gas stream, although tolerated, had reduced both (a) the efficiency of the conversion of alkanes to synthesis gas and (b) the life of these catalysts.

Having reviewed the NGC Gas Contract with Amoco, I was persuaded that the more pertinent long-term issue to be resolved was not why the slug catcher at Galeota Point had failed but how to supply gas continuously to petrochemical producers at Point Lisas that was satisfactory for reformer catalysts. Accordingly, I broadened the terms of reference of the committee to investigate the nature of natural gas that entered into NGC's pipeline system and how the optimum gas quality could be delivered to consumers at Point Lisas.

In April 1985, the Technical Manager at NGC and I visited all suitable offshore locations by helicopter, and gas samples were taken. These samples were subsequently analysed at Chromaspec, a Houston subsidiary of Core Laboratory, to determine the dew points and compositional analyses of the natural gas at those locations. In parallel, with the agreement of Mr.

Robert Powers, the then President of Amoco, I was permitted to use Amoco's computer facilities in the Tatil building to plot dew point curves based upon the gas composition of the gas samples that were taken offshore.

The Problem. The phase diagrams showed that the hydrocarbons generally existed in two phases at the offshore locations, as illustrated by the analysis of a sample of the exit gas from the 3rd stage knock-out vessel, which was taken on the NGC Poui platform. The slug catcher facilities at Galeota Point, when in operation, had collected the liquid accumulations that had arisen in the offshore pipelines due to subsea cooling, isothermally, and the drop in pressure after leaving the offshore platforms. Onshore, as the gas was transmitted along the 24' pipeline between Galeota Point and Picton, with the further drop in pressure, more of the heavier hydrocarbons left the gaseous phase and entered the liquid phase under the ambient isothermal conditions on land. Thus, the drip pot at Picton was always full. This situation continued along the pipeline from Picton to Point Lisas, where, eventually, NGC delivered to its customers "wet gas" consisting of methane and heavier alkanes.

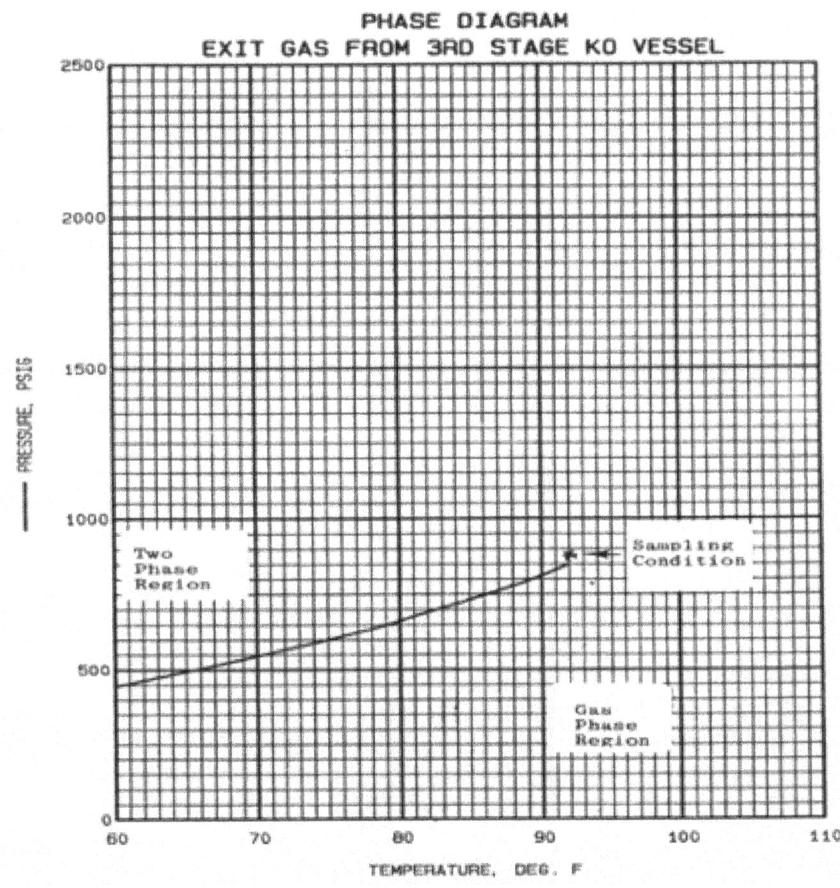

In September 1995, I submitted my report to the NEC and NGC under cover of a confidential memo (below), which led to the formation of Phoenix Park Gas Processors Limited:

"It was with the background as a part-time lecturer on the properties of petroleum reservoir fluids at the University of the West Indies (U.W.I) that the problem of liquid in natural gas was technically approached. Having taught the whole curriculum on gas properties at U.W.I. for three years, both reason and intuition directed my approach to the problem, a solution for which seemed conventional, particularly after examining some of the clauses in the gas contract (Article XI: 11.2, Attachment 7 of the·report). If Cassia continued to be the main supplier, there seemed only one way to go and the question, "who was at fault?" bore less relevance in the context of clauses 8.3 and 8.4 of the contract. The real issue was, therefore, strategy. Mindful of the veiled motives, I nevertheless took advantage of the AMOCO invitation to use their computer facilities to study the technical aspect of the problem, which also facilitated discussions with expatriate staff on a less formal basis and provided a deeper insight to the mood of the negotiations and ATOC. These experiences, therefore, were such as to recommend continued links with the Company, which, if properly utilised could result in significant benefits for Trinidad.

As Chairman of the committee, I submitted the interim and draft reports to senior management of the National Gas Company (NGC) and National Energy Corporation (NEC) for inputs, after which the draft was adjusted by committee members of the NEC and AMOCO and assembled finally at two locations: at Point Lisas, the NEC facilities were used by NEC personnel on the committee, and another report which also contained the legal information was bound by the author under strict cover at TATIL Building and hand-delivered to relevant management. Later, only the less-sensitive copies were presented at a meeting of the full committee without any discussion of the contractual details of gas quality in the presence of AMOCO. I have little doubt that the contents of the report reflect the feelings of the committee, but more so the cause of the problem.

One of the recommendations - the removal of certain hydrocarbon components from the gas stream - has the potential to be a foreign exchange earner and can provide strategic supplies of LPG and condensate at a time when the refinery at Pointe-a-Pierre may need economic support. It may therefore be more beneficial to the economy if a natural gas liquid recovery study, for example, markets, site location, economics, etc., were explored immediately with a view to implementing a decision in the New Year. The contents of the report by the present committee could be examined in total as a parallel exercise, but the urgent need for winning foreign exchange for Trinidad and Tobago may show greater relevance.

There are two conceptual approaches to a solution to this phase of the problem. In the first instance, the economics would be based on maximum liquid recovery and consequential improvement in the quality of gas supply to reformer catalysts. In the other, the natural gas liquid recovery would be determined by the tolerance levels of catalysts to heavy hydrocarbons in the gas stream, which though conservative, will have a lesser impact on the sales volumes and BTU content of the resulting gas stream. The choice of process is, therefore, critical since costs, yields, and other cost factors will assist the final process selection.

At the outset, the markets look encouraging. For example, in July and August 1985, 243,000 bbls and 320,000 bbls of condensate were imported, respectively, for MOGAS blending, perhaps the start of a trend because of the incentive in utilising naphtha streams preferentially in the hydrogenation to higher valued components. Condensate sells for about \$27-28 (U.S)/bbl whereas platformate, at 1.5¢ - 2¢ per U.S.G. lower than the posted price of MOGAS 95, currently fetches 83¢ per U.S.G. (\$33.20/bbl). For LPG, the market opportunities are even more secure. January - June sales for 1985 were 236,053 bbls (average 1304 BOPD), and local demand in August of 41,199 bbls has been growing annually at about 7%, which could easily be supplied by cooling the Galeota stream (appendix 2). The price of LPG is fixed by GORTT at 40¢ (TT)/lb, and the export market price is tied to the MOGAS 83 price of 78¢ U.S./USG (\$1.89 TT), but any substitution by LPG from natural gas into the local market still means freeing up of an equivalent supply for the more lucrative export market.

A conservative approach, i.e., to select a process to meet the needs of Point Lisas catalysts, is expected to daily generate significant revenue from the LPG stream alone if gas demand is projected at 450 MM scfd.

The future looks promising particularly because most of the infrastructure is in place to carry out additional studies locally (appendix E), and as Chairman of the Committee to Investigate the Problem of Liquid in Natural Gas on behalf of the NGC, I thank those responsible for giving me the opportunity to serve the citizens of Trinidad and Tobago.

DR. R. L. MONTEIL
COMMITTEE CHAIRMAN
Attachments:

1985:09:16.

Downstream Natural Gas-Based Petrochemicals. In 1984, when I joined the NEC, the sole methanol plant, the Trinidad and Tobago Methanol Company, was owned by the NEC. Methanol is produced by reforming natural gas with steam to synthesis gas ($CO + H_2$) followed by the high-pressure catalytic conversion of the synthesis gas to create pure methanol.

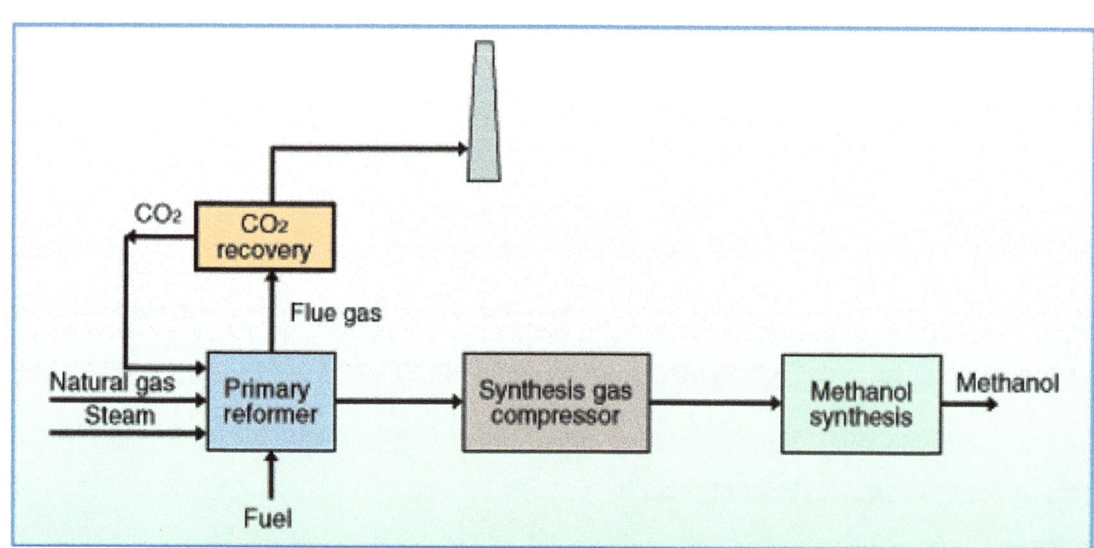

Steam-methane reforming reaction: $CH_4 + H_2O$ (+heat) $\rightarrow$ [$CO + 3H_2$] (synthesis gas)

Catalytic conversion reaction to methanol: $CO + 3H_2 \rightarrow CH_3OH$ (methanol)

Ammonia is manufactured by using purified Hydrogen that is obtained from synthesis gas and then treating Hydrogen with Nitrogen obtained from the atmosphere.

The synthesis of ammonia: $3H_2 + N_2 \rightarrow 2NH_3$ (ammonia)

At the NEC, one of my functions was to develop value-added industries, further down the value chain from the primary petrochemicals, that would contribute to the growth and development of the national economy. Key commodity chemicals from natural gas (in blue) are shown below.

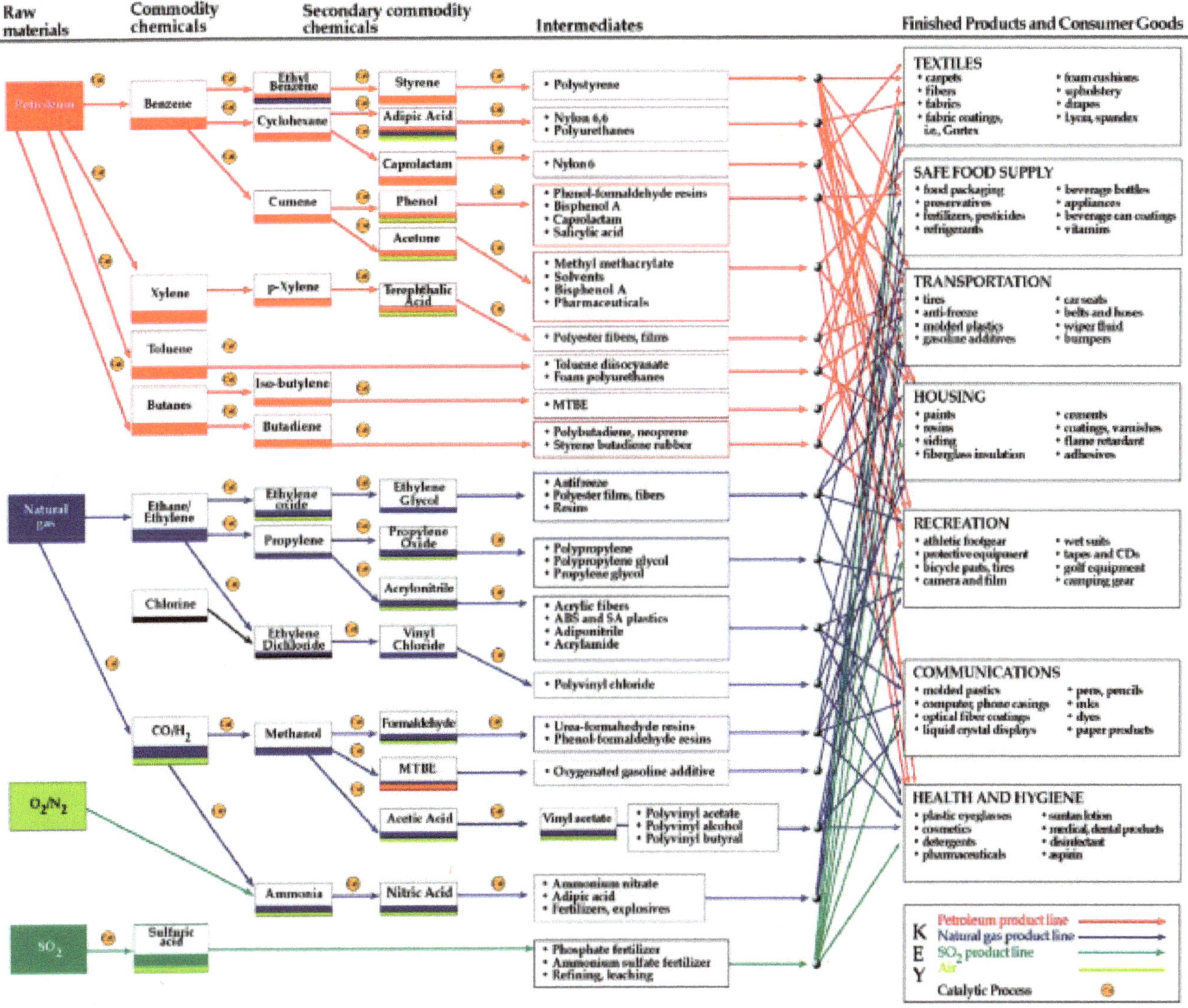

In 1984, as illustrated, in Trinidad and Tobago, there were two potential sources of feedstock for downstream petrochemical manufacture locally:

(a) Methanol and ammonia from primary natural gas-based petrochemicals and

(b) Benzene, toluene, and xylene, the "BTX," from the Texaco refinery streams (if that refinery fractionally separated the BTX streams).

The petrochemicals, which are based upon methane from natural gas ("C_1- based petrochemicals"), end up in two markets: fertilizer and polymeric adhesives, whose leading outlets are in construction materials. On the other hand, ethylene and benzene, traditionally obtained from oil refineries, are used largely as feedstock for high-value polymers such as the leading plastics, fibres, and rubber.

By September 1985, I had established that there were licensors of technology who were interested in equity participation in various downstream C_1-based petrochemicals projects in Trinidad and Tobago. The main projects were the manufacture of acetic acid, melamine, and methylamines. Other potential investors had proposed using natural gas to produce C_2 intermediates, a potentially lucrative approach, but these projects had not yet been fully commercialised.

Acetic Acid. In 1985, most of acetic acid produced in the world had been consumed in the manufacture of vinyl acetate (40%), which in turn, had dominant end-use markets in adhesives (35%), paints (20%), paper, and textile coatings (20%). Ethylene and acetic acid were traditional feedstocks for vinyl acetate monomer ("VAM"). During my review in the 1980s, market growth for acetic acid was estimated at 3.7 % for production based upon methanol.

Commercial Syntheses. Acetic acid was commercially produced by one of three routes: methanol carbonylation, butane oxidation, and acetaldehyde oxidation. Synthetic acetic acid plants had capacities between 35-275 million Kg/year and the most appropriate method of synthesis of acetic acid was by the Monsanto process, which was based upon methanol carbonylation, using precious metal catalysts like rhodium. The less efficient acetaldehyde-based process had represented 25% of U.S. capacity and 41% of Western European capacity.

The Monsanto Process (methanol carbonylation)

CH_3OH (methanol) + CO (carbon monoxide) $\rightarrow$ CH_3COOH (acetic acid)

A comparison of the economics for acetic acid production from different feedstocks showed that the total production costs (U.S. cents/pound) for methanol carbonylation, n-butane oxidation, and acetaldehyde oxidation were 11.85, 24.73 and 20.22, respectively. Thus, methanol carbonylation was clearly the most economical route.

After Monsanto introduced the low-pressure methanol carbonylation process for acetic acid production, industry acceptance was very rapid. The process had provided nearly 100% conversion at almost 99% selectivity to acetic acid, and the last three plants that had been built had used that technology. Since there was significant value-added in converting methanol to acetic acid, I sought joint venture opportunities for constructing an acetic acid plant in Trinidad to satisfy a potential market of 674 MTPY using methanol as feedstock.

The Proposals. Monsanto confirmed to me that they were willing to licence their process to manufacture methanol in Trinidad and that to be competitive in the world acetic acid commodity market, a normal plant capacity would be 150,000 metric tonnes per year (MTPY) and cost U.S.$ 37 million based upon U.S. Gulf Coast conditions. Smaller plants would have been justified if low-cost carbon monoxide or low-cost natural gas were available and if we had captive markets for acetic acid production. A 60,000 MTPY plant would have cost about U.S.$23 million.

Monsanto was also interested in equity participation and had written me to advise that if a CO_2 source were to be available, installing a cold box and recovering carbon monoxide from the synthesis gas reformer would cost approximately 5 million dollars. It was feasible to integrate a CO-separation purification unit with the existing methanol plant and produce a 40-60K MT acetic acid plant from methanol. Further, the economics of an acetic acid plant would be improved,

> "… if *the NEC is planning to build a second methanol plant, the integration of a carbon monoxide supply when building the plant would have been the least expensive alternative.*"

Halcon Technology. I also received an alternative proposal for acetic acid production using Halcon SD Group technology, which also utilised methanol and carbon monoxide. The Halcon technology was capable of either producing acetic acid alone or co-producing acetic acid with acetic anhydride and vinyl acetate. At the time, the majority of acetic anhydride consumption had been used to manufacture cellulose acetate flake, which had an end use in cigarette filters. However, local vinyl acetate production was unnecessary in the absence of a source of ethylene.

C$_2$-based Petrochemicals. Traditionally, the naphtha and other streams from a refinery were used to produce the feedstock for petrochemicals based upon two carbon atoms. Benzene, for example, could be cracked to produce ethylene and ethylene was the building block for many secondary commodity chemicals. Alternatively, natural gas could also be used to produce petrochemicals based on two carbon atoms such as ethanol (CH_3CH_2OH), the most promising C2 project.

On 12th June 1985, Halcon wrote to me expressing interest in ethanol production:

"Dear Dr. Monteil:

I am writing to follow up [our] letter to you of March 26, 1985, which discussed the Halcon SD ethanol technology. This process uses methanol and natural gas as feedstocks to produce ethanol suitable for addition to gasoline as an octane enhancer and volume extender.

Ethanol produced in Trinidad and Tobago from indigenous raw materials could be the lowest cost ethanol in the world and could be exported to the U.S. duty free under the Caribbean Basin Initiative. We believe that it would compete successfully with fermentation ethanol produced within the U.S. and with imports from other regions.

Ethanol production is a logical way for Trinidad and Tobago to participate in the markets for octane enhancers in the U.S., Canada, Western Europe, and Japan; these continue to grow as lead is removed from gasoline.

To achieve minimum cost, the ethanol plant requirements for methanol and synthesis gas should be incorporated into the design of the methanol plant. For example, if 1000 tons/day of ethanol is added to the 1500 tons/day of methanol already planned, total methanol production will grow to about 2200 tons/day. This will place the methanol plant among the largest and most economical in the world, providing low cost methanol for marketing and for conversion to ethanol ......

The Halcon SD ethanol process is flexible and can be used to produce ethanol alone or ethanol with one or more co-product acetyls. These include acetic acid, acetic anhydride, and ethyl acetate...."

Second only to water as a solvent, ethanol had been employed in nearly all industries, with the major demand being in toiletries, cosmetics, pharmaceuticals, and a wide range of chemical applications (including acetic acid and ethylene glycol). Production of ethylene glycol had ranked No. 28 in the top 50 chemicals in the U.S. in 1982 and 1983, owing to the great demand for anti-freeze and polyesters, its two major markets. Ethanol, therefore, was an attractive prospect.

Methylamines. Methylamines are made by the reaction of methanol and ammonia (which were locally available) in a continuous flow system in the presence of a dehydration catalyst. The reaction produces a mixture of mono-, di-, and tri-substituted methylamines that are separated by extractive distillation[7] after unreacted raw materials have been removed. The three methylamines are formed in proportions that depend upon the chosen reaction conditions.

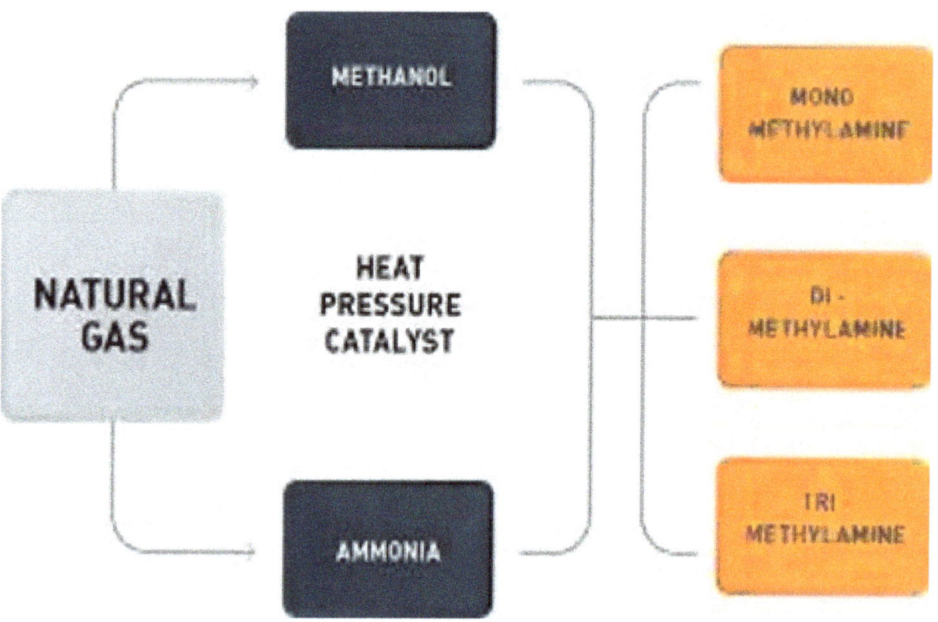

Plant Sizes. In the 1980s, "Leonard Technology" was used to produce three-quarters of the world's demand for methylamines. Plants varied in size from 10 million pounds/year, predominantly in developing countries, to 100 million pounds/year in developed Nations. The Leonard process could have been used to produce only methylamines or adapted to a multi-unit plant producing a variety of amines such as methyl-, ethyl-, and iso-propylamines.

Capital Costs. The capital investment for a battery-limits plant that produced the three amines (with plant capacities between 10-100 pounds/year at a Gulf Coast location) ranged from U.S.\$2 million to U.S.\$14 million (at 1981 prices). The low capital costs made methylamines an attractive prospect particularly since both raw materials (methanol and ammonia) were available. Included in the battery-limits were the processing units, 15 days' raw materials, product storage, and technical costs such as licence fees and design, purchasing, procurement, and construction costs.

---

[7] Extractive distillation is a means of separating two or more substances with similar boiling points by adding another substance that changes the boiling point of one of compounds, thus making them easier to separate.

Proposals. In February 1985, I wrote to various companies, including Union Carbide ("UCB"), a licensor of technology for the manufacture of methylamines. Although I received a response from UCB's licensing division fairly quickly, confirming its willingness to license its technology, it appeared that increasing negative publicity due to the catastrophic Bhopal incident in India had delayed subsequent progress on a proposal.[8]

Uses. In 1985, over one-half of the world consumption of methylamine was in the manufacture of "Sevin" insecticide, and the other half had surfactants as a major outlet. About 50% of dimethylamine had been used to manufacture dimethylformamide and dimethylacetamide, which are solvents that are used during the manufacture of acrylic fibres. Other large outlets were for surfactants and for accelerators for the rubber industry. Trimethylamine was used exclusively in the manufacture of choline chloride, an animal feed supplement.

Melamine. Melamine is a white, crystalline powder produced from urea. In 1985, more than 90% of melamine production worldwide was used as a raw material to make the liquid resin amino-formaldehyde, which was then processed and used in a variety of applications, including laminates. These laminates, by imparting excellent resistance to heat, water and most chemicals, exhibit good electrical properties and surface hardness.

Melamine   Formaldehyde   M-F prepolymer

---

[8] The Bhopal incident occurred when methyl isocyanate, which is produced from methylamine, and is an intermediate chemical in the production of carbamate pesticides, was released into the atmosphere eventually killing many people.

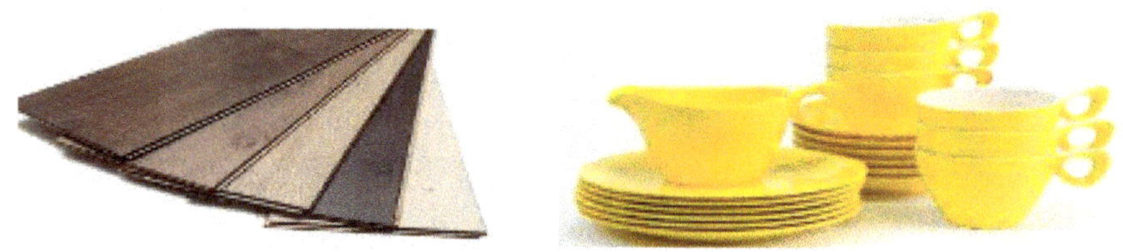

Markets. In 1985, the end-use of melamine resins in adhesives for plywood and fiberglass represented the fastest-growing market in the U.S. Urea-formaldehyde resins, which were also used in the building industry, were relatively easily produced from urea and formaldehyde (obtained by the oxidation of methanol).

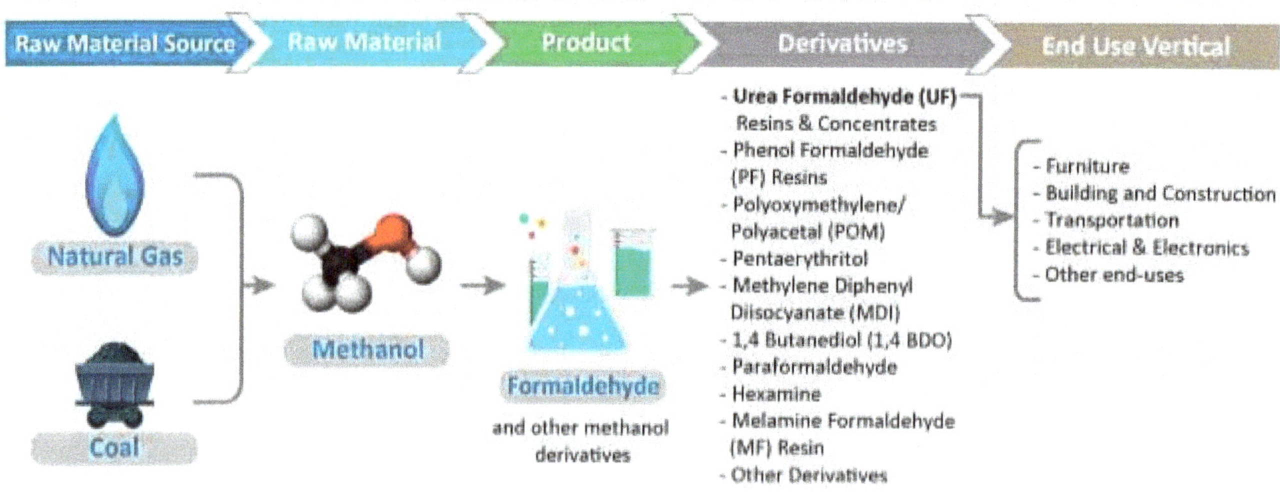

On 26th September 1985, I submitted a report of my work on the prospects for further downstream natural gas based industries, including active proposals and relevant discussions with respect to Acetic Acid, Melamine, and Methylamines.

For family reasons, the beginning of the 1985 academic year, the 1st week in October, was the most appropriate time for me to leave Trinidad to pursue legal studies in England, and I requested a leave of absence from the NEC. However, after a discussion with the CEO, Mr. Eldon Warner, who was very encouraging about my plans, I resigned because I did not satisfy the necessary criteria for being granted a leave of absence.

Chapter 3: A Legal Interlude

With my previous academic qualifications, I was able to register at City University, London for the 1985-6 Diploma-in-law, which was the academic part ("Part 1") of the legal training for the Bar in England and Wales. On successful completion of the course and graduation in August 1996, I attended the vocational part ("Part 2") of the training for the Bar at the Inns of Court School of Law in the 1986-87 Academic Year.

I sat the Bar examinations in the Great Hall of Lincoln's Inn, which, with my apprehension, could easily have been Newgate prison, where prisoners used to be hanged. In any event, I passed the Bar exam with 2nd class honours, and that day, my wife Michele, and I thanked the Highest Authority in the Inner Temple chapel. In July 1987, I was called to the Bar of Middle Temple.

Having been called to the Bar, I undertook pupillage[9] in intellectual property law (patents, trademarks, passing off, and copyright) at specialist chambers in Inner Temple.

---

[9] Akin to an apprenticeship where a "pupil" functions under the watchful eye of an experienced lawyer, who is referred to as the "pupil master". In practice, the pupil will review a brief and offer an opinion to the pupil master.

After almost two years at the Bar, I joined the Treasury Solicitor's department, which conducted litigation on behalf of, or provided legal advice to, other Government departments. My office was located at "Queen's Anne Chambers," which was opposite to St. James' Park underground station and very close to St James' Park.

I was almost exclusively involved in company law litigation, including actions brought by the Monopolies and Mergers Commission under section 7 of the Fair Trading Act, 1973. The 1973 Act had consolidated and extended U.K. competition law by controlling Monopolies, Mergers and Takeovers, Restrictive Trade Agreements, and Resale Prices.

Confirmation. My interview for confirmation in the Treasury Solicitor's department one year later was memorable because of a question asked by the Home Office representative on the interview panel. Noting my chemistry background, he had asked my opinion about a case where defendants had been charged with setting off explosives, and the main evidence against them was the presence of nitroglycerine on their fingers. The defendants had claimed that they were playing with cards, which, apparently, are known to contain traces of nitroglycerine. The State had convicted the defendants mainly based on forensic evidence that had used thin layer chromatography ("TLC") to identify the nitroglycerine.

At the interview, when asked, I indicated that TLC was normally used as a qualitative and **not** a quantitative tool for the identification of chemicals. Thus, I explained that I would use TLC to follow the course of a reaction to establish whether starting materials were still present and, accordingly, it would have been more appropriate to have used a quantitative analytical method such as gas chromatography. I referred the panel to analytical chemists at Imperial College, London. The legal significance of the question that I was asked is that the State can petition the Court of Appeal (which it did) to have convictions quashed if the convictions were subsequently perceived to be "unsafe or unsatisfactory." The lasting impression on me was the desire for the Crown not to "win the case" but to pursue justice in the public interest.

Energy Law. After my confirmation at the Treasury Solicitor, I was appointed an advisory lawyer to the Department of Energy. My advisory work involved several aspects of the wide and varied U.K. energy legislation but mainly legal issues that arose due to exploration and production activities on the U.K. Continental Shelf ("UKCS"). My new office building was located at "1 Palace Street," and my room overlooked trees on a remote part of the Buckingham Palace compound. Later, a more senior colleague pulled rank, and I no longer had 'a room with a view'.

The office security was strict. Special files were kept in a vault that was secured by a steel door with a combination lock, and the filing cabinets in each person's room also had combination locks. In our absence at night, the security would examine offices to see whether documents had been put away and the filing cabinets in their rooms had been secured. If not, as I soon learnt, the security would transfer any papers that had been left on my desk and place them in the filing cabinet, which would then be clamped and secured with a long metal bar. Anyone who has received the 'Denver Boot' or a ticket from a traffic warden would appreciate my reaction the following morning. In any event, the only practical recourse was to go to security and fill out a report, after which the metal bar on the filing cabinet would be removed.

At the time, pursuant to the *European Communities Act, 1972*, the domestic legislative environment was subject to E.U. Directives and their interpretation as determined by the European Court of Justice.[10] However, for policy reasons, in energy matters, the application of the Treaty of Rome to the UKCS was generally not conceded when providing legal advice. The genesis of the policy was primarily to ensure that the benefits of North Sea Oil, in particular the benefits of refining, accrued to the U.K. In order to implement this policy, a provision had been incorporated into all offshore production licenses that UKCS-produced petroleum should be landed first in the U.K. unless the Secretary of State had consented otherwise.

---

[10] The European Communities Act 1972 brought the U.K. into the European Union and the Act also incorporated Community Law into the domestic law of the United Kingdom.

In the early stages, Her Majesty's Government ("HMG") had avoided confrontation with the European Commission over the application of Community Directives to offshore oil and gas activities (and others concerning environmental matters). However, the situation changed when environmental pressure groups began to take an increasing interest in the award of hydrocarbon licences as exploitation activities moved closer to the shore and as more environmentally sensitive blocks became available.

I was generally inundated with requests for advice relating to the transfer of interests among licensees, which involved painstaking scrutiny of bundles of legal documents to ensure that HMG's interests were protected. Once, the Secretary of State came close to exercising his powers under section 23 of the Petroleum and Submarine Pipelines Act, 1975, which gave the Secretary of State power to permit third-party use of an owner's pipeline in certain circumstances. Ultimately, the parties came to an agreement, presumably, having recognised that it was not in their interests for the Secretary of State to know the entire commercial details of their operations.

There are a few legal matters that I remember very well.

1. The Arms to Iraq Affair. Following the first Gulf War of 1991, resulting from the Iraqi invasion of neighbouring Kuwait, there was interest in the extent to which British companies had been supplying Saddam Hussein's regime with weapons and materials used to prosecute the war. HMG had been accused of endorsing sales of arms by U.K. companies to Iraq during a period when there was a U.N. embargo on such sales.

The pertinent legal issues had involved the copious use of the infrequently deployed Public Interest Immunity ("PII") Certificates. This is a procedure whereby information alleged to be highly sensitive to "national security," is withheld from public disclosure. The certificates were issued by relevant Government ministers based upon the legal advice of the Attorney General and Andrew Leithhead in the Treasury Solicitor's Department. Andrew Leithhead was my former supervisor in the Treasury Solicitor's Department, and I functioned under his guidance whenever I was involved in a matter where the State commenced litigation in the public interest. When the 'Arms to Iraq' scandal exploded, I wondered whether Andrew (who was very honest but gentle) had been unfairly treated in the press and, later, by the subsequent Scott Report. The details of this cause célèbre, therefore, have held a special significance for me.

The Law. The right to a fair trial is predicated on the principle of equality of arms, of which an essential element is the defendant's right to have access to the information collected by the prosecution during an investigation. Disclosure of the evidence is determined by a single test:

> *Any material that might reasonably be considered capable of undermining the prosecution case or assisting the case for the defence should be disclosed.*

However, the right to disclosure of this material is limited by the doctrine of PII, which allows the court to withhold relevant information from the defence when the court decides that it is in the public interest to do so.[11] The determinative role of the court was at the core of the legal advice that had been provided by the Attorney General and the Treasury Solicitor's department.

The Context. Matrix Churchill was a Coventry firm involved in exporting to Iraq, machine tools, which could be used for making military equipment. Three senior executives of Matrix Churchill were charged in 1991 with deceiving the Government over the intended use of the machine tools when they applied for export licences. The defence of the executives was twofold:

(a) that Alan Clark, as a junior minister at the DTI, had, in 1988, encouraged them to conceal any military use and to emphasise the civilian applications of the equipment; and

(b) that in any event the Government had known of the true intended use, not least because one of the defendants (Paul Henderson) had been a regular contact of the intelligence agencies.

To establish these defences, access was sought to a wide range of documents covering inter-departmental discussions and advice to ministers on licensing applications for the export of machine tools to Iraq, and documents from the intelligence agencies. However, four Government ministers, including Michael Heseltine, sought to prevent disclosure by PII certificates. Michael

---

[11] *Conway v Rimmer* [1968] A.C. 910.

Heseltine had signed his certificate but reluctantly, having been advised to do so. His certificate did, however, make plain that he was doing no more than asserting that the documents fell within the requisite classes, and he had been advised that the last word lay with the trial judge to balance the public interest. Ultimately, the case was abandoned following the evidence of Alan Clark, who had admitted under cross-examination that notes of one DTI meeting were "economical with the actualité," and he knew full well the machine tools could be used to manufacture munitions.

The case endorsed a critical lesson for me - the importance of providing reasoned, transparent and **independent** advice, which I have adopted throughout my professional career.

2. The Gas Pipeline Interconnector Treaty between the U.K. and Ireland. In Europe, the gas was provided primarily by three suppliers external to the EU: Norway's Statoil, Russia's Gazprom, and Algeria's Sonatrach. Gazprom and Sonatrach were State companies. Thus, a source of concern and debate associated with the projected increase of gas imports from Russia, the Middle East, and Africa had been related to the political reliability of these countries where nationalistic policies or possible internal instability were perceived as major threats. Additionally, a major fear was that oil and gas producers, together with their national Governments, could increasingly link their export policies to political considerations using Europe's dependence on natural gas as a tool of political pressure. Indeed, natural gas, being pipeline-bound, could be easily suspended by action in the host country.

Thus, European policymakers had focused on preventing politically-based natural gas supply disruptions, which had emanated mainly from Moscow's use of the "energy weapon." Ultimately, proposals were approved for the creation of a single European energy market that had been largely motivated by a desire to reduce Europe's dependence on Russian gas at a time of tension over a conflict in Eastern Ukraine. These proposals also addressed the development of trans-European networks for transporting electricity and gas. In that light, the Government of the Republic of Ireland ("Eire") decided in December 1991 that Bord Gáis Éireann (BGE) should undertake the construction of an undersea pipeline connecting the natural gas grids of Eire and the U.K.

Accordingly, the Government of the U.K. and the Government of the Eire entered into an agreement relating to the transmission of natural gas by a pipeline ("the Interconnector") from the natural gas custody transfer point at the British Gas Compressor Station at Beattock near Moffat in Scotland to the pig trap at the Irish Gas Board Shore Station at Ballustree near Loughshinny in Eire.

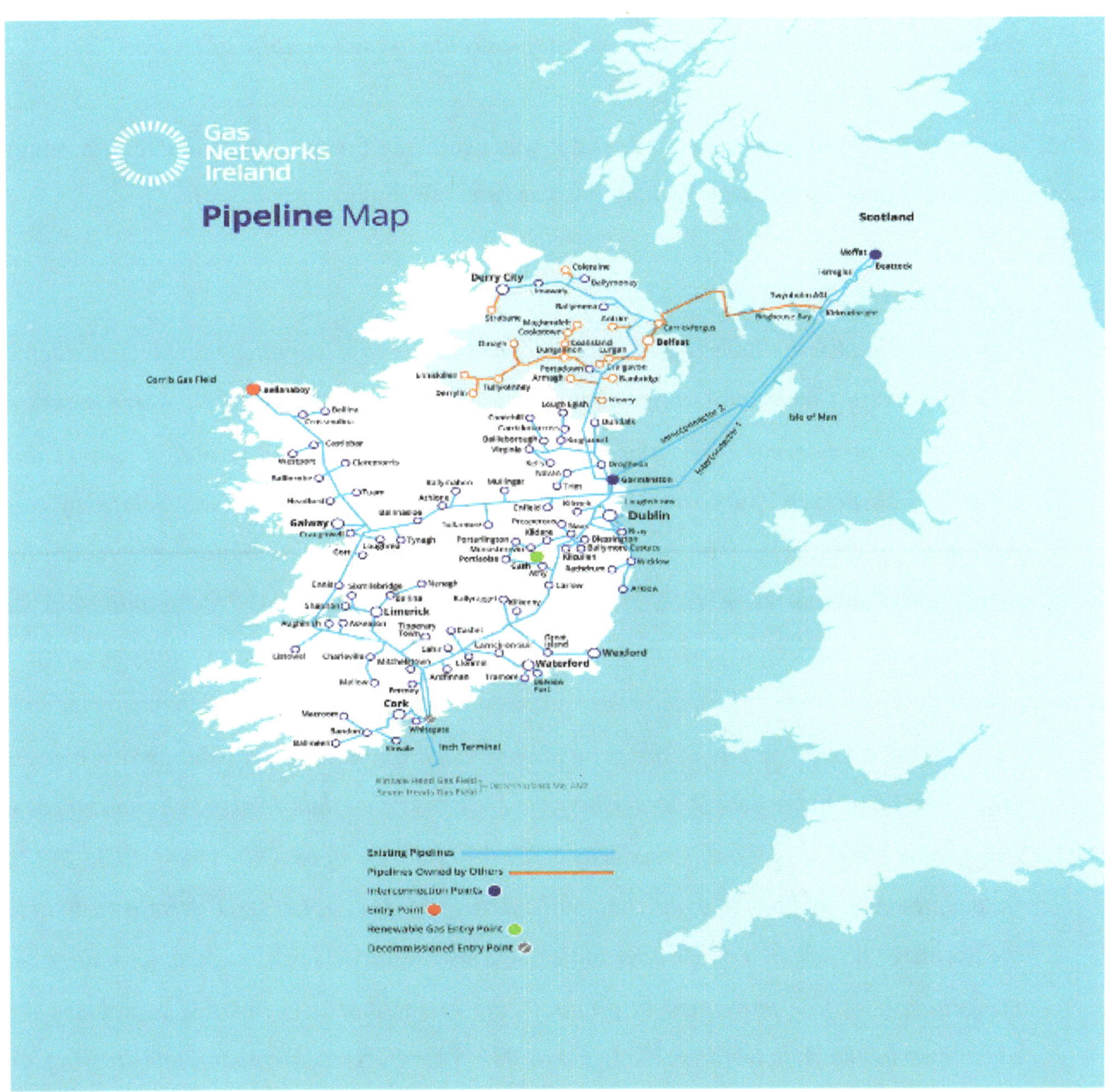

The Treaty governing the Interconnector was the first of its kind between two Member States of the European Union. I was on the team that negotiated the Interconnector Treaty as the U.K.'s legal representative on energy matters, and the Treaty was signed on 30th April 1993. Later, I had to consider many residual issues relating to the Interconnector Treaty, including the following question when the Irish authorities argued that HMG had no authority to take the gas that would be flowing to them through the Interconnector:

> *"In an emergency situation in Great Britain, could the Secretary of State lawfully use his powers under the Energy Act to utilise the gas that had originated from a field in Norway and was flowing through the Interconnector en route to the Republic of Ireland?"*

More 'heat' than light had been generated, and it was necessary for me to provide very thorough reasoning in order to quell the issue. My advice was in the following terms:

"The Treaty of Rome

Any powers under the Energy Act to restrict or prohibit gas supplies must be considered by reference to Community obligations (section 2(4) of the ECA 1972), which makes it appropriate to determine first the extent of the U.K.'s Community obligations in this context. For present purposes, the most relevant general provisions relate to free movement of goods in Articles 30-34 of the EEC Treaty (elimination of quantitative restrictions between Member States) and Article 36, which permits "prohibitions or restrictions on imports, exports or goods in transit .." on the grounds of public security.

*"Security of supply" is one of the grounds on which quantitative restrictions on exports may be justified under Article 36 because the ECJ has ruled that, in some circumstances it can be an element of "public security" (Case 72/83 Campus Oil [1984] ECR 2727). Indeed, it was stated by the court that whilst Article 36 does not reserve certain matters to the exclusive jurisdiction of Member States, it allows national legislation to derogate from the principle of free movement of goods to the extent that it is justified to achieve the objectives in the Article (at p.2751, para. 32). Therefore, a Member State, relying on Article 36, can impose a prohibition on a particular export **and** satisfy community obligations provided any exercise of power to prohibit that export goes no further than is required for "public security." The treaty of Rome, therefore, does **not** prevent the*

Secretary of State from exercising his emergency powers in respect of gas transiting the U.K. en route to Ireland.

The Gas Transit Directive

As far as the completion of the internal market for energy is concerned, more particularly in the natural gas sector, this is to be achieved by the Gas Transit Directive (91/296/EEC) and by a proposed Directive (discussed later) concerning the common rules for the internal market. Notably, Article 3 of the Gas Transit Directive expressly allows a Member State to take into account its security of supply ("conditions of transit shall ... not endanger security of supply nor quality of service ..."). Accordingly, if, pursuant to section 3(1)(b) of the Energy Act 1976, it is declared that there exists or is imminent in the United Kingdom an actual or threatened emergency affecting fuel or electricity supplies, the Secretary of State has the power to give directions to prohibit or restrict the supply anywhere in the world of natural gas to specified persons (see section 2(2)(b)).

Thus, the proposition, insofar as it relates to gas transit, is equally misconceived because the Secretary of State's emergency powers can be exercised in relation to a gas transit contract without conflict with his Community obligations.

International law on goods in Transit

It is not entirely clear what legal argument would apply for the Irish authorities if the gas is the property of other persons before the custody transfer point. In particular, if the Secretary of State were to exercise Energy Act powers in respect of natural gas which originated from outside U.K. jurisdiction and entered the U.K. (albeit "in transit"), I am puzzled why any persons should claim that the U.K. had breached international obligations to them in respect of gas which at no time had become their property. Surely it is the owner of the gas who may wish to argue that he had had his goods commandeered?

*Insofar as an argument is founded upon Article V of GATT (which lays down rules on freedom of transit), SIOT v. Ministero Delle Finanze [1983] ECR 731 clearly confirms that the GATT rules "do **not** [my emphasis] apply to transit within the Community" (p.781* *para. 31) and "govern only the Community's relations with the other contracting parties and cannot be applied within the community itself" (p.776 para. 12). Again, the proposition is misconceived because when natural gas is "in transit" in the U.K., the relevant international obligation is similar to that in respect of gas originating within U.K. jurisdiction, i.e., Article 36 of the EEC treaty.*

Discretion

*The Secretary of State's power to prohibit or restrict the supply of natural gas also relates to the supplies of gas to the Republic of Ireland from a U.K. producer or aggregator. However, any exercise of discretion in respect of such gas must be in the context of the circumstances prevailing at the time when the Secretary of State is contemplating the exercise of his power. In particular, it would be an unlawful fetter on his discretion if he were to purport to give effect to an interpretation of the legislation (wider than his Community Obligations) that specific supplies of gas would **always** be excluded from the provisions of the Energy Act. The proper course is to make an appropriate amendment to the Energy Act by primary legislation if, as a policy consideration, the Secretary of State wishes to circumscribe his powers by exempting certain supplies of natural gas.*

In this context, the only guidance I am able to give is that the underlying policy when the Act came into force was that HMG must be able to move quickly to protect the national interest when faced with unpredictable and quick changes in the supply situation.

The Interconnector Treaty

The Government may negotiate and conclude any terms of a treaty, and in this context, I can do no better than cite the views of Lord Templeman in International Tin Council [1989] 3 W.L.R. 969 (at 980 G):

> *'A treaty is a contract between the governments of two or more sovereign states. International law regulates the relations between sovereign states and determines the validity, the interpretation, and the enforcement of treaties. A treaty to which Her Majesty's Government is a party does not alter the laws of the United Kingdom. A treaty may be incorporated into and alter the laws of the United Kingdom by*

means of legislation. Except to the extent that a treaty becomes incorporated into the laws of the United Kingdom by statute, the courts of the United Kingdom have no power to enforce treaty rights and obligations at the behest of a sovereign government or at the behest of a private individual.'

*Accordingly, a Treaty obligation (which is not incorporated into domestic law) would not "override" domestic legislation, although it would be contrary to agreed principles for a country to enter into an international obligation and **then** attempt to use the absence of the obligation in the domestic legislation as a means of not complying with that obligation. This stems from the Vienna Convention on the law of treaties (1980):*

> *'Every treaty in force is binding upon the parties to it and must be performed by them in good faith (Article 26)' and 'A party may not invoke the provisions of its internal law as justification for its failure to perform a treaty (Article 27)'.*

Therefore, if the Secretary of State were to make a commitment in the Interconnector Pipeline Treaty that Energy Act powers would not be used in respect of certain supplies of natural gas, it must be given on the understanding that domestic legislation would be appropriately amended to give effect to that obligation.

The Equality Principle

Should there be a disruption in supply in a neighbouring Community country, it is to be expected that the market forces of the internal energy market will come into play. In this context, one can anticipate cooperation between Member States. For example, certain precautionary measures have been taken at Community level to deal with difficulties in supplies of crude oil and petroleum products (but not gas), and Member States are required to maintain minimum stocks and to co-ordinate the national measures adopted for the purpose of drawing on those stocks.

However, as far as I am aware, there is no principle established by virtue of an existing obligation that Member States should, in the event of a disruption of supply, take measures to ensure that their respective markets receive __equal__ treatment. As to Article 23 of the proposed directive, this permits a Member State to "take the necessary protection measures" save that such measures must cause the least possible disturbance in the functioning of the internal market and must not be wider in scope than is strictly necessary to remedy the sudden difficulties.

Clearly, the Commission, without deciding on the circumstances, has considered that there may be situations when a Member State could justifiably take protection measures.

Moreover, Article 23 should not be taken in isolation but should be read in conjunction with other proposed provisions. For example, Articles 12 and 19 (obligations to transmit or distribute gas) reflect the ambulatory provision that Member States should determine the extent and nature of the rights and public service obligations of the distribution company. Those obligations include the obligation ".. to supply, the maintenance of system security and the development of the system capability to meet demand ..," which are matters that may well show variation in Member States.

The relevant obligation in Article 23 above, namely, that protective measures should cause the least possible disturbance in the internal market, is not inconsistent with the obligation under Article 36 supra, namely, that protection measures should not go beyond that which is required for "public security." Those obligations do not extend as far as requiring a Member State to ensure that in taking protective measures in an emergency, another Member State is to be treated equally.

*Previously a proposal had been sought by the Irish authorities in the treaty negotiations that the U.K. would not use its emergency powers "**even if a force majeure situation exists in its territory, including a state of war, a national gas or energy shortage or similar emergency.**"*

While such a proposal could not have been accepted, as a good community neighbour, there was something to be said for both Governments agreeing to consult each other with a view to establishing the procedure to the adopted in the event of a disruption in supply in either territory. That, it seemed to me, was as far as we could have gone in the absence

of a decision to amend the Energy Act."

The conclusions in the last paragraph above were reflected in Article 11 of the Interconnector Treaty (Co-operation in the Event of Disruption of Supply), where both parties had agreed:

"The two Governments, recognising each other's legitimate interest in safeguarding supplies of natural gas to consumers and in maintaining system security and capability, shall consult each other, at the earliest opportunity after the entry into force of this Agreement, in order to establish the framework for co-operation in the event of a serious disruption in natural gas supplies."

3. Abandonment of Offshore Installations. Under Article 2.1 of the Convention on the Continental Shelf done at Geneva on 29th April 1958 (the "1958 Convention"), the coastal State exercises, over the Continental Shelf, sovereign rights for the purpose of exploring it and exploiting its natural resources. The U.K. is a party to the 1958 Convention, and the treaty came into force on 10th June 1964. By the Continental Shelf Act 1964, the U.K. gave effect to certain provisions of the 1958 Convention. Thus, by section 1 of the 1964 Act -

"Any rights exercisable by the United Kingdom outside territorial waters with respect to the sea bed and subsoil and their natural resources, except so far as they are exercisable in relation to coal, became vested in Her Majesty."

By Article 5 of the 1958 Convention, the exploration of the Continental Shelf and the exploitation of its natural resources *"must not result in any unjustifiable interference with navigation, fishing or the conservation of the living resources of the sea."* Further, any *"installations which are abandoned or disused must be entirely removed."*

The need for the removal of outdated platforms and installations had become urgent since the 1980s due to (a) technological advances in the oil and gas industries and (b) the exploration of resources to new depths and distances (which had increased the number of structures in the sea and the impacts on the seabed).

The Petroleum Act 1987 (1987 c. 12), which received Royal Assent on 9th April 1987, gave effect to the U.K.'s obligations under international law. The Act had made provision for the abandonment of installations and submarine pipelines, and the responsibility for their removal rested with the owners of the installation or pipeline.

The Issue. The question arose for my consideration in a situation where the owner of an installation was either no longer in existence or had abandoned an installation in the North Sea:

> *"What liability, if any, to pay compensation would fall upon HMG if there were damage or loss of life caused by fishing gear becoming entangled with pipelines that were not removed in accordance with an abandonment plan previously approved by the Secretary of State?"*

After a lot of drama, I provided the advice below:

> *"In the context of international obligations, the issue is the extent of the rights and duties of the United Kingdom with respect to the decommissioning of submarine pipelines and what responsibility accrues. Liability will accrue to the United Kingdom if it breaches any international obligation (by omission as well as by act), considered, as a general proposition, by reference to customary international law. It would be open to a foreign State (who has, for example, lost a state-owned ship) to bring a claim under international law before an international court or tribunal alleging State responsibility under international law on the part of the U.K. Sovereign immunity would not be available as a defence in these circumstances.*
>
> *Article 2 of the 1958 Geneva Convention on the Continental Shelf (GCCS), which is regarded to express customary international law, confers sovereign rights upon the coastal state for the purpose of exploring the Continental Shelf and exploiting its natural resources. However, by Article 5.1, the exploration of the Continental Shelf and the*

exploitation of natural resources, "must not result in <u>unjustifiable interference</u> with navigation, fishing or the conservation of the living resources of the sea...."

Thus, it would appear that some interference with, for example, fishing, is permitted provided it is not "unjustifiable" (whatever that means).

Article 60(3) of the 1982 United Nations Convention on the Law of the Sea (UNCLOS)[12] refers to "installations and structures" which, if abandoned or disused, "shall be removed to ensure the safety of navigation, taking into account any generally accepted international standards established in this regard by the competent international organisation. Such removal shall also have due regard to fishing, the protection of the marine environment, and the rights and duties of other States." The 1982 Convention has not been ratified by the U.K.,[13] but some writers consider it to reflect customary international law.

Although both the GCCS and UNCLOS Conventions contain clauses relating to the abandonment of offshore structures, neither Convention deals explicitly with the question of pipelines. The position for pipelines is, therefore, unclear. Thus, some academics have concluded, on the basis of Article 5.1 of the 1958 Convention, that there would be no duty to remove pipelines; but once decommissioned, they must be rendered in such a condition that they cannot unreasonably interfere with legitimate uses of the sea.

Based on that interpretation, the international obligation would be not to permit the pipelines to remain in circumstances where they are potentially capable of interfering with or even jeopardising other States' fishing activities on the High Seas. That view is consistent with advice given by [others] in the context of a requirement for partial as opposed to complete removal of installations. However, if Article 60(3) of UNCLOS is considered to be customary international law, arguments can be advanced that the

---

[12] The United Nations Convention on the Law of the Sea lays down a comprehensive regime of law and order in the world's oceans and seas establishing rules governing all uses of the oceans and their resources. The Convention was opened for signature on 10th December 1982 in Montego Bay, Jamaica and it came into force in accordance with its article 308 on 16th November 1994, 12 months after the date of deposit of the sixtieth instrument of ratification or accession. Today, it is the globally recognised regime dealing with all matters relating to the law of the sea.

[13] It was ratified subsequently by the U.K on the 25th July 1997 and took effect on 24th August 1997.

obligation - "installations or structures which are abandoned or disused shall be removed to ensure safety .. taking into account generally accepted international standards .." - should cover pipelines. Although a "structure," it seems, is something other than an installation, why, in principle, should there be a distinction as to what 'litter' may be left on the seabed?

Notably, Article 60(3) of UNCLOS appears to contemplate a basic rule of removal, with something being less possible by reference to certain international standards.

Other academic opinions suggest that pipelines are to be understood as being covered by the international law provisions on installations and structures by virtue of the argument that permitted installations and structures are those necessary for the exploration and exploitation of the Continental Shelf. If, as it is argued, pipelines come into this category, once they become unnecessary, a duty to <u>remove</u> them arises.

Liability. *The principles for State liability were set out in the leading case of <u>Corfu Channel</u> (U.K. v Albania, decided by the ICJ in 1949). The Court held that there was an obligation to give warning of dangers in the territorial sea, that Albania failed to give the warning even though it must have known about the minefield, and that damage to RN ships resulted from the failure to give a warning. A State that had complied with requirements to give a warning and had complied with standards established by the IMO, would be in a position to deny liability on the grounds that a violation of an international obligation could not be demonstrated. Accordingly, if claims were brought by a foreign State against HMG for compensation in respect of loss or damage caused to the former's interest due to fishing gear becoming entangled with pipelines, our defence would be based on the proposition that international law does not require entire removal in all circumstances and that what had been done by the U.K. in the instant case was reasonable, taking into account the legitimate interests of all users of the waters.*

Of course, if the relevant international obligation is not to cause "unjustifiable interference with fishing," arguments that the obligation has not been discharged, become stronger when the pipelines are no longer buried. This situation might arise, as I understood from our discussion, with a constant scraping of the seabed (by legitimate fishing). Much would depend on the precise facts of the case, but clearly, the risk of successful claims would be reduced by giving publicity as to the depths, position, and dimensions of the pipelines.

Any decision of an International Tribunal such as the ICJ would depend, I think, on the current state of the international law at the time.

Ownership*. The problems relating to ownership can arise in situations other than implied in the question, i.e., when the entity is no longer present having ceased to trade. If the pipelines have been left with the apparent intention of abandonment, possession and ownership would not <u>necessarily</u> cease, but there is room for argument. For example, the long-standing principle of English Common law - that "a man cannot relinquish property he has to his goods unless they be vested in another" (<u>Haynes Case</u> (1613) 12 Co. Rep. 113) - has been excepted in a number of English cases relating to wrecks of vessels. But it would appear from the "wreck" cases that the owner would still remain subject to a duty to protect other vessels from receiving injury from the wreck.*

It would nevertheless be entirely speculative to suggest what obligations would be left on an owner of pipelines that had been abandoned in accordance with the requirements of the Secretary of State pursuant to the Petroleum Act. Any owner would probably argue that he did what he was required to do; if that were proved to be unsatisfactory, the blame, I think, would then fall upon the Secretary of State. I understand that we have little past experience in these matters, but it would be relevant (but not necessarily dispositive) as to what the owners of previously abandoned pipelines were required to do.

In any event, if it is not appropriate or possible in these particular circumstances to cause the company to incur further costs, it would be prudent to obtain 'clearance' from the relevant fishing authorities.

The position is more complicated if, as envisaged in your minute, the [entity] were no longer in existence. But an assumption that there will be "no one to sue" begs the question if there are circumstances where HMG can be said to have ownership of the pipelines. For example, under section 25(1) of the Petroleum and Submarine Pipelines Act 1975, when a works authorisation ceases to be in force, the controlled pipeline to which it relates shall, if it has not been removed, be transferred to and vest in the Secretary of State free from encumbrances. Also, a works authorisation that does not contain terms providing for it to be of unlimited duration ceases to be in force on the expiration of any period of its duration specified under the terms of the authorisation (see section 24(1)(a)). It will therefore be necessary to see the terms of the authorisation before coming to conclusions in this regard.

There is also authority for the proposition that goods that are "lost, abandoned or ownerless" <u>may</u> fall to the Crown under the rule 'quod nullus est fit domini regis'. Moreover, section 2(1) of the Crown Proceedings Act 1947 established the general principle that the Crown is subject to the same liability in tort as if it were a private person in respect of any breach of the common law duties of an owner of property.

It would therefore be sensible to make a cautious assumption that an aggrieved person will find HMG an attractive target, particularly if large sums of money were involved. If HMG were regarded to be the owner of the pipelines, a respectable argument can be made that the common law position would apply. As one judge put it in a wreck case:

> *'If I own a dog which I know bites mankind and it bites a person, I do not escape liability to that person by saying: I abandon the dog.'*

Accordingly, a person who has lost property or suffered personal injury by fishing gear becoming entangled with pipelines could conceivably contemplate an action in negligence against the Secretary of State. Such an action might be based on the approval of the burial (as opposed to the removal) of the pipelines and/or a failure to remove them.

__The Forum__. However, whether or not any action is likely to be successful will depend on several considerations, including the law of the country where the proceedings are brought. In the U.K., one would need to know the particulars of negligence to form a view as to the prospects of success in any such action. I suspect that much will depend on whether, in the given case, there was "unjustifiable interference" with the legitimate rights of other persons. Whilst that is entirely a question of fact, it is easy, at the outset, to make a distinction between an accident caused by, on the one hand, pipelines that are reasonably buried and, on the other, where they are not. In either case, the argument for an element of contributory negligence by the person who has suffered damage would be enhanced if the Secretary of State had taken the necessary warning steps.

Finally, the House of Commons Energy Committee, in the 14th Report of its 1990/91 session, recommended (paragraph 32) that

> *"... an acceptable low risk of debris is presented by a buried or trenched pipeline in a stable section of the seabed. However, .. sections of pipelines resting on, buried or trenched in unstable sections of the seabed, should be removed to avert the danger of debris or obstruction to sea traffic."*

Are we considering a <u>stable</u> section of the seabed? In any event, you should not assume that the burial of the pipelines is the end of the matter: what might be appropriate in 1994 to satisfy international obligations might not be adequate in the future. You should also bear in mind that if pipelines are buried, they are more easily forgotten. How would one know if the pipelines, on a future occasion, are at the level of the sea bed?"

4. Deepwater Dumping in International Waters. The International Council for the Exploration of the Sea reported in 1968 that huge amounts of waste were being discharged into the North Sea. On a Norwegian initiative, the Nordic Ministers of Foreign Affairs decided that the Nordic countries should jointly approach all members of the North-East Atlantic Fisheries Convention to discuss the matter and to take action. In the opinion of the Nordic countries, there was conclusive evidence that increased dumping of substances injurious to marine life was taking place. An absolute ban on dumping, as proposed by the Nordic countries, went too far in the opinion of the other participating countries. Thus, in order to come to a common viewpoint, the parties met on 23rd July 1971 in Paris.

A few days before the meeting in Paris, on 16th July 1971, the coaster *"Stella Maris"* left the port of Rotterdam to dump 650 tonnes of chlorinated hydrocarbons in the northern part of the North Sea. Initially, the intention was to dump the waste at about 180 nautical miles off the Norwegian coast, but after Dutch Government involvement, a dumping site was chosen a thousand miles from both the Icelandic and Irish coasts.

The waste in question consisted of 5% dichlorethane, 50% trichlorethane, and the rest consisted of tar products. The company concerned had the intention of making regular trips to dump this waste, which had resulted from the production of vinyl chloride.

As a result of protests from the Governments of both Iceland and Eire against the dumping at the new spot, on 25th July, the ship returned to Rotterdam and unloaded its cargo in order to avoid further negative publicity.

The Conventions. These events led to the adoption of the Oslo Convention on 15th February 1972 in Oslo, Norway, which was designed to control the dumping of harmful substances from ships and aircraft into the sea. The Oslo Convention came into force on 7th April 1974, and the United Kingdom became a signatory to this Convention in 1975. On 4th June 1974, a Convention for the prevention of marine pollution from land-based sources was concluded in Paris (the "Paris Convention") and was ratified by the United Kingdom on 6th April 1978. In Paris, on 22nd September 1992, the Convention for the Protection of the Marine Environment of the North-East Atlantic was open for signature at the Ministerial Meeting of the Oslo and Paris Commissions (the "OSPAR Convention"). The OSPAR Convention replaced the Oslo and Paris Conventions when it came into force on 25th March 1998.

The Issue. In February 1994, prior to the U.K.'s adoption of the OSPAR Convention, a somewhat innocuous matter came to me for legal advice concerning the disposal of a floating oil storage buoy in the North Sea, the "Brent Spar". My advice supported environmental rather than economic decision-making, mercifully, because the matter later exploded with significant international disquiet.[14]

The Brent Spar had entered service in 1976 as a storage stop for crude oil from the nearby Brent field, but it ceased operations in 1991 when a new pipeline connecting the drilling platforms directly to the mainland rendered the structure obsolete. After examining various options and carrying out a risk assessment and an environmental impact assessment, Shell, the operator, decided to sink Brent Spar in the deep ocean. The disposal options had been the subject of negotiations between Shell and HMG for more than three years, and in December 1994, HMG approved the plans for deep water sinking of the Brent Spar.

---

[14] https://nvdatabase.swarthmore.edu/content/greenpeace-campaigns-against-dumping-brent-spar-oil-rig-1995.

The preferred sinking solution involved towing the rig to deep water 150 miles to the west of Scotland, breaking it up with explosives, and sinking it, along with the residual oil, sludges, and waste products remaining in its tanks. It would have been one of the first installations in the North Sea to reach the end of its life and be decommissioned in this way.

The Dispute. In my briefing, I was told that the installation had contained light fittings, a low specific activity scale (LSA), and a residual sludge from offshore petroleum production consisting of compounds based upon heavy metals, including those derived from Mercury and Cadmium. Further, the Ministry of Agriculture, Food and Fisheries ("MAFF") had said that in accordance with the provisions of the 1972 Oslo Convention, the Contracting Parties to the Convention must be consulted before the Installation containing those substances is abandoned. On the other hand, DTI officials had believed that such consultation was not intended under provisional guidelines that were adopted in 1991 and that consultation, in any event, would cause unnecessary delay.

At that time, the 1992 OSPAR Convention had not been ratified by the U.K. Thus, I had given my advice using the 1972 Oslo Convention and its interpretation and guidelines to determine the U.K.'s then-current international obligations for the protection of the marine environment. And, since it had seemed to me that the DTI officials had perceived, wrongly, that there was no need for consultation, I had considered it appropriate to provide highly technical and detailed reasoning for my advice (which later proved to be fortuitous). The bizarre proposal would have been more easily understood if my advice were now read in conjunction with the Annexes that are referenced.

*"**Trace Quantities**. The dumping of substances listed in Annex1 to the 1972 Convention ("Annex 1 substances"), such as Mercury and Cadmium and their compounds, is prohibited under Article 5 of the Convention. However, by Article 8(2), the provisions of Article 5 would not apply if these substances 'occur as trace contaminants in waste to which they have not been added for the purpose of being dumped.' Assuming that the Cadmium and Mercury substances have not been "added" for the purpose of dumping, the first issue is whether, as a question of fact, those substances would be considered to be present in "trace" quantities by reference to the relevant guidelines. In any event, such dumping would still 'remain subject to Articles 6 and 7' (Article 8(2)), and a specific permit would be required from the 'appropriate national authority' by reference to the provisions in Annexes II and III.*

The guidelines for interpreting the terms used in the Convention are contained in section A7 of the Manual, and paragraph 2 relates to the definitions of the terms used. It is said, under "trace contaminants," that the definition would effectively be determined by whether or not a waste containing Annex 1 substances successfully passed the test procedures associated with the Prior Consultation Procedure (PCP). The associated test procedures on the PCP (set out in Appendix 1 of Chapter 3 of the Manual) contain specific concentration limits for inorganic Mercury and inorganic Cadmium. In particular, where the respective concentrations of Mercury and Cadmium exceed 0.3 mg/litre and 1.0 mg/litre, respectively, the waste must be subject to the PCP.

However, the total concentration in the waste should <u>never</u> exceed 1.0 mg/litre inorganic Mercury and 4.5 mg/litre of inorganic Cadmium (see paragraph 2). These are dissolved concentrations, unlike the concentrations of Mercury and Cadmium in the sludge, which are set out in Table 4 of the Metocean report (2 ppm and 58 ppm, respectively). A correlation should be made, bearing in mind that the Oslo Commission observed at its twelfth meeting that sludges with particulate concentrations of Mercury and Cadmium are expected to be far higher than dissolved concentrations. From 5(i) of your minute, the prognosis does not appear to be good.

If the total concentration of either Mercury or Cadmium exceeded the prescribed limit, it follows that the substance would not then be present in "trace" quantities under the relevant guidelines. In such a situation, I doubt whether arguments to the effect that there is no interference with fishing (Annex III general considerations) would apply to the exclusion of the prohibition in Article 5 against dumping Annex 1 substances. In any event, under paragraph 4 of Annex II, dumping would only be allowed in deep water where (a) the depth is not less than 2,000 metres and (b) the distance from the nearest land is not less than 150 nautical miles. The conclusion in 5(iv) of your minute appears to be accurate.

The PCP. *A prior consultation procedure for the dumping of wastes containing Annex 1 substances was first adopted by the Oslo Commission at its first meeting in Oslo in 1974, and this principle was reaffirmed at its 7th meeting in Brussels in 1981. At the 9th meeting in Berlin in 1983, the revised prior consultation procedure made it a requirement for the contracting party proposing to issue a permit to "inform the Secretary of the Oslo Commission as soon as possible but no later than three months prior to the time of the envisaged dumping operation" (paragraph 3).*

In paragraph 5, it is stated that 'the Secretary shall forward the information submitted by the contracting party immediately to all contracting parties.' This is consistent with the flow chart in appendix 3 to the procedures and decisions manual; for concentrations of mercury greater than 0.3 mg/litre (or 1.0 mg/litre cadmium), evaluation of all aspects of the waste should be made, and the prior consultation procedure must be followed.

The 1991 guidelines relate to the disposal of offshore installations and do not appear to address the question of consultation in respect of Annex 1 substances. However, paragraph 1.1 of part A prohibits a Contracting Party from disposing at sea, offshore installations containing Annex 1 substances unless those substances can be exempted as trace contaminants under Article 8(2). Clearly, there must still be some mechanism for the determination of what constitutes a "trace" contaminant. Thus, in the absence of an express provision to the contrary, the presumption, in my view, is that the 1983 guidelines still apply to make it a requirement for consultation in respect of the dumping of any Annex

I substances present in the installation. Accordingly, the MAFF observations in paragraph 5(ii) of your minute seem reasonable.

***Prior Notification.** At its 10th meeting in Oslo in 1984, the Commission adopted a prior notification procedure ("PNP"), which was to be followed by a contracting party proposing to issue a specific permit under Article 6. The objective of the PNP is to enable other contracting parties to give advice to the country issuing a specific permit on alternative disposal options that might be available. The Commission agreed that a period of sixty days after the notification of the specific permit to the secretariat should be allowed before the dumping operation would commence. At its 12th meeting in Madrid in 1986, the Commission also agreed that "no exception should be made for the prior notification procedure if the relevant criteria are fulfilled" (paragraph 4 of the PNP).*

***Annex II substances.** The heavy metals present in the sludge also include lead, copper, zinc, and arsenic, which are Annex II substances, the dumping of "significant" quantities of which requires, under Article 6, a specific permit from the appropriate national authority. From the interpretation provisions, a substance would be regarded to be present in "significant" quantities if it constitutes more than 0.1% of the weight of the quantity of waste for disposal. It would appear from Table 3 of the report that the quantities of zinc and copper are "significant," but you will need to check this."*

The Epilogue. When Greenpeace discovered the dumping plans for the Brent Spar in the Atlantic, it went into action, it has been suggested,[15] in order to avoid a precedent for future dumping of toxic substances into marine environments. Thus, Greenpeace's *Moby Dick* landed at the Brent Spar on 30th April 1995, and climbers occupied the rig while *Moby Dick* had stood by.

---

[15]   'Greenpeace campaigns against dumping the Brent Spar oil rig, 1995', https://nvdatabase.swarthmore.edu/content/greenpeace-campaigns-against-dumping-brent-spar-oil-rig-1995.

Clean up
your mess,
Shell!
GREENPEACE

GREENPEACE

By 4th May 1995, many Greenpeace activists occupied the Brent Spar and conducted live TV interviews using satellite equipment. That action generated wide publicity around Europe.

By 16th May, opposition parties in the U.K. denounced the dumping, and by the 17th May, Belgium and Iceland had joined Denmark in condemning the British Government for allowing the dumping. Germany added its voice to growing international pressure on 24th May. Later, European nations at the OSPAR Commission agreed to a ban on the dumping of offshore steel oil rigs.

The protest of Greenpeace against the deep-sea disposal led to a major consumer boycott against Shell, and within weeks, the company suffered a significant loss of market share in Central Europe. On 20th June, after over two months of mounting protest, worsening publicity, and growing public pressure, Shell stopped its ocean disposal plan for the Brent Spar. In 1997, an independent commission assessed eight disposal options and listed dumping as the worst environmental option. Shell later produced a dossier,[16] which in its introduction, indicates:

---

[16] Brent Spar Dossier, https://www.shell.co.uk/sustainability/decommissioning/brent-spar-dossier/_jcr_content/par/textimage.stream/1426853000847/32a2d94fa77c57684b3cad7d06bf6c7b65473faa/brent-spar-dossier.pdf

Finally, in 1998, Shell announced on-land recycling of the Brent Spar in Norway, with plans to use scrap for foundations of a new ferry terminal.[17] The Brent Spar was towed ashore.

5. Pre-Article 169 Letter to the European Commission. From time to time, various interest groups would complain to the European Commission, alleging HMG's failure to implement various Community Directives that governed operations in the oil and gas sector. After such complaints, the Commission, through a "pre-Article 169 Letter", would request the U.K. to submit

---

[17] *Ibid*. After evaluation of disposal options, a re-use option was considered to have "the best all-round environmental benefits" with an estimated cost of £23-£26 million. The deep sea disposal option was assessed to have "fewer environmental benefits but no significant environmental impacts" with an estimated cost of £17-£20 million.

information in order to determine whether there had been non-compliance with the Directives before a formal decision was made under the provisions of Article 169 of the EC Treaty:

"If the Commission considers that a Member State has failed to fulfil an obligation under this Treaty, it shall deliver a reasoned opinion on the matter after giving the State concerned the opportunity to submit its observations."

On Tuesday, 28th February 1995, I submitted my final legal advice to colleagues, which concerned a pre-169 Letter as the result of a challenge to the entire petroleum licensing regime in the U.K. The complainant had alleged that the U.K. had failed to implement the Habitats -, Freedom of Information -, and Environmental Impact Assessment - Directives. My legal advice had to address (and tiptoe around) many sensitive policy issues within the framework of the ECJ jurisprudence:

(a) To what extent could administrative practices be acceptable in lieu of legislation?

(b) Did the EIA Directive create rights for individuals, and if so, had the U.K. given effect to those rights in sufficiently clear and precise terms?

(c) To what extent did the EIA Directive apply offshore, particularly with respect to activity on the UKCS beyond a twenty-five mile band around the coast?

(d) Should Exploratory Drilling be subject to an EIA?

(e) Were there economic considerations that would justify overriding the obligations in *Habitats?*

(f) Would the licensing procedures in the 16th Round comply with the spirit of *Habitats*?

(g) Should the conditions attached to a licence be disclosed pursuant to the Freedom of Information Directive?

I do not know the outcome of the pre-Article 169 Letter because the following day, Ash Wednesday, 1st March 1995, just a few months short of 10 years after leaving the NEC, I left England and took up the post as LNG Coordinator at NGC on Thursday 2nd March 1995.

Chapter 4: Atlantic LNG Train 1

Why LNG? During the 1970s, Trinidad and Tobago used natural gas, principally for electricity generation and ammonia manufacture. By 1995, NGC's total gas consumption had grown to almost 700 MMscfd due to an increase in gas utilisation for power generation and additional plants at Point Lisas. The projections were that the proven reserves would last for at least 30 years more, extended to 55 years, if the "discounted probable" reserves were considered. Thus, Atlantic LNG Train 1 could have been comfortably accommodated on the basis of proven reserves alone, despite needing more than twice as much gas as NGC's largest existing customer.

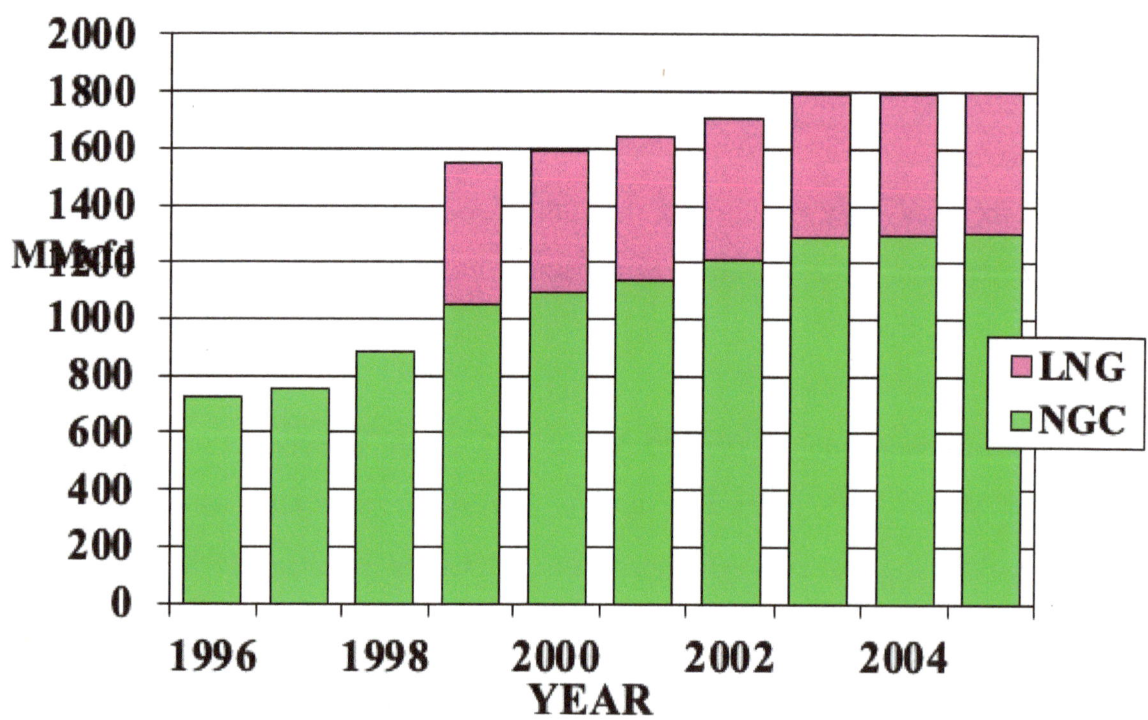

The doubters who insisted that a single train LNG plant could not be built on a greenfield site in southwestern Trinidad for under U.S.$1bn, were finally silenced when the construction, testing and work came to an end. The 1st LNG Train was then formally handed over to its owner, Atlantic LNG Company of Trinidad and Tobago (Atlantic 1), by the contractor, Overseas Bechtel.[1] "Unique," said Simon Bonini, then president of BG Trinidad, to describe the achievement:

> *"The project lacked the economies of scale the engineers said was vital if we were to get our unit costs down to a reasonable level. We were going against all conventional LNG wisdom. Venezuela was expected to beat us to the punch. Yet, we pulled it off."*[2]

It is also reported[3] that Martin Houston, then head of British Gas PLC's Trinidad operations, recalls a competitor commenting:

> *"If you build an LNG facility in Trinidad, I will eat my shoes."*

Two years later, at a ground-breaking ceremony, Martin Houston suggested to the crowd that the skeptic be sent *"a little bit of hot pepper sauce for his shoes."*

LNG Records. Not only did the five shareholders, Amoco Trinidad (LNG) BV (34%), British Gas Trinidad LNG (26%), Repsol LNG Port of Spain BV (20%), Cabot Trinidad LNG Ltd (10%), and NGC Trinidad and Tobago LNG (10%) – pull it off, they established world LNG records[4]:

- Atlantic 1 had become the world's largest single-train LNG facility, involving the input of 164 bcf of gas a year and 3 million tonnes a year of LNG output;
- It was the world's lowest cost single train LNG facility at U.S.$950m;

---

[1] 'Trinidad sets World Standards', Energy Day, June 1999, http://static.conocophillips.com/files/resources/smid_016_trinidad_sets_world_standards.pdf (assessed on 6th February 2022).

[2] *Ibid.*

[3] **How Trinidad Became a Big Supplier Of Liquefied Natural Gas to the U.S.'**, *Alexei Barrionuevo, The Wall Street Journal*, March 13, 2001, https://www.wsj.com/articles/SB98443474293628576.

[4] 'Trinidad sets World Standards', Energy Day, June 1999, *supra.*

- It was the world's fastest LNG project, taking six-and-a-half years from conception to completion, including 34 months for construction and handover;
- It had used ground-breaking payment arrangements based on netback from sales revenue in the U.S. and Spanish markets;
- It had been built on innovative interlocking relationships among a gas supplier, plant owners, and LNG buyers, all of whom had specific interests along the value chain;
- It had used first-time financial arrangements with bankers for an LNG investment, including no pledge of shares, liberal ownership transfer covenants, plant expansion without lender approval under certain circumstances, and monthly dividend payments;
- It was the only LNG export plant in entire Latin America and the Caribbean region and only the second LNG plant to be built in the entire Western Hemisphere; and
- It had demonstrated that an LNG project could cost less (without being bigger than existing plants) and could also supply non-traditional markets while gaining project financing.

How did it happen? Much of the credit should go to Gordon Shearer, a geophysicist born and raised in Scotland, and Dr. Kenneth Julien, an electrical engineer, by profession, at the University of the West Indies. The pair, having won over Patrick Manning, then the Prime Minister of Trinidad and Tobago, entered into a Memorandum of Understanding ("MOU") on 16th December 1992. The objectives of the MOU were (among other things):

(a) *"To export approximately 300 MMscfd of natural gas, as LNG, from Trinidad to the Everett LNG terminal in Boston, Massachusetts, U.S.A.;*

(b) *To study the economic and technical feasibility of designing, financing, constructing, and operating a liquefaction plant with a capacity of approximately 300 MMscfd;*

(c) *To study the forms of commercial and contractual structures which were necessary to support the development of the project, including the gas supply, gas transportation, construction, operating and maintenance, LNG supply, and shipping contracts; and*

(d) *The establishment of a project schedule for achieving those objectives."*

The MOU expressed, in good faith, the intentions of NGC and Cabot based on their respective understandings and the information available to each of them, and it was signed by Ken Julien and Gordon Shearer on behalf of their respective companies:

Dr. Kenneth S. Julien
Chairman
For and on behalf of
The National Gas Company of
Trinidad and Tobago

R. Gordon Shearer
President
For and on behalf of
Cabot LNG Corporation

ON 11TH JANUARY 1993, BG ACCEDED TO THE MOU AS AN ADDITIONAL PARTY.

Dear Sirs,

RE: LIQUEFIED NATURAL GAS PROJECT

The National Gas Company of Trinidad and Tobago Limited ("NGC") and Cabot LNG Corporation ("Cabot") have entered into a Memorandum of Understanding dated December 16, 1992 regarding the possibility of developing a project to export approximately 300 million cubic feet of liquefied natural gas ("LNG") per day from Trinidad to the Everett, Massachusetts LNG terminal owned by a subsidiary of Cabot and to other possible markets as may be identified by the parties. A copy of the Memorandum of Understanding is attached as Appendix 1.

BG Trinidad, Inc ("BGT") has expressed its desire to associate itself with NGC and Cabot in the study and possible development of this project, and NGC and Cabot are agreeable to BGT acceding to the Memorandum of Understanding as an additional party. Accordingly, NGC and Cabot hereby agree that BGT shall, upon execution of this accession letter in

/...2

----

the space provided below, become an additional party to the Memorandum of Understanding, subject to all of the terms and conditions thereof, it being understood that nothing in this accession letter or the Memorandum of Understanding creates any partnership, joint venture, or other legal association or relationship among the parties or gives rise to any rights, duties or binding obligations among them, except for the obligation of confidentiality specified in the Memorandum of Understanding.

Malcolm A. Jones
Managing Director
For and on behalf of
The National Gas Company of
Trinidad and Tobago Limited

R. Gordon Shearer
President
For and on behalf of
Cabot LNG Corporation

Agreed and assented to

Peter F. Jessen
Project Director, Trinidad
For and on behalf of
BG Trinidad, Inc

Apparently, Amoco was more difficult to persuade. In Ken Julien's words, as reported,[5] they had to be brought "screaming to the table." According to the Wall Street Journal,[6] the company's chairman and chief executive, Larry Fuller, was worried that even if LNG demand in the U.S. were strong enough, Trinidad would not have enough gas to satisfy the demand.

---

[5] 'How Trinidad Became a Big Supplier Of Liquefied Natural Gas to the U.S.', *Alexei Barrionuevo, The Wall Street Journal,* March 13, 2001, https://www.wsj.com/articles/SB98443474293628576.
[6] *Ibid.*

Even a strong personal connection between Cabot's Chairman, Sam Bodman, and Mr. Fuller was insufficient to sway Mr. Fuller. However, it was not until the latter part of 1993, when Dr. Julien and Prime Minister Manning visited Amoco's Chicago headquarters, that they were able to convince Mr. Fuller, despite his reservations. Additionally, in December 1993, Mr. Manning wrote Sr. J. Baldosa, the Chairman of the Board of Enagas, notifying him of the project and inviting him to visit Trinidad and Tobago.

"December 14, 1993
Sr. J. Badosa
Chairman of the Board
ENAGAS SA
Avenida de America 38
Madrid
SPAIN.

Dear Sr. Badosa ··

TRINIDAD AND TOBAGO LNG PROJECT

I would like to take this opportunity to write to you about the proposed liquefied natural gas (LNG) project in Trinidad and Tobago.

My government has taken a number of important initiatives to support the LNG export project, which is being developed by Amoco, British Gas, Cabot, and the National Gas Company of Trinidad and Tobago (NGC). Senior representatives of the project are visiting your company shortly to establish your interest in purchasing LNG from Trinidad.

This project is the most important energy-related venture currently being developed in Trinidad and Tobago. My government has taken and will take all steps necessary to facilitate and expedite the project in a timely manner. In this respect, the Government has approved the utilisation of proven gas reserves off the East Coast of Trinidad. The Government is also developing a large industrial zone with an existing deep-water harbour of which the LNG plant will be the major feature.

Trinidad and Tobago has a long history of stable democratic government and a hospitable investment climate. Potential buyers such as yourselves can therefore be confident that LNG from Trinidad will represent a reliable source of supply well into the next century.

I would like to extend my personal invitation to you to visit Trinidad and Tobago so that you can fully appreciate the reasons why we believe LNG from Trinidad will be the next source of supplies in the Atlantic Basin... ."

The LNG Markets. The LNG markets in New England and Spain were critical factors for decision-making by the project sponsors. The salient features of these markets had been furnished to a number of banks and institutions to assist them in their due diligence for project financing.

Worldwide, the majority of LNG was exported to the Far East, and although LNG accounted for a mere 6% of European gas supply, it was the dominant source of gas supply in Spain (81%) and a critical source of gas supply in Belgium (38%) and France (27%). LNG, while accounting for a smaller volume of trade than gas delivered by pipeline, had shown a greater growth rate than gas transported via pipeline. As early as 1995, LNG growth had been anticipated globally.

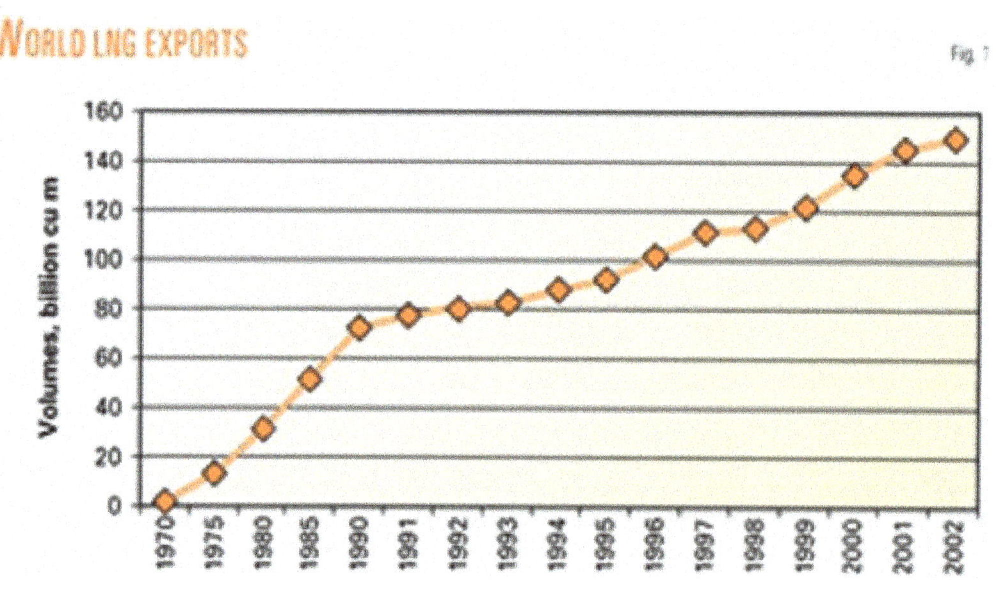

Source: Oil and Gas Journal, June 23, 2003

Gas Supply Security. Importers of LNG in Europe and the U.S. had mainly relied upon the traditional exporters, Algeria and Libya, for LNG supply but were increasingly seeking to reduce their dependence on these major LNG exporting countries and diversify their gas supply.

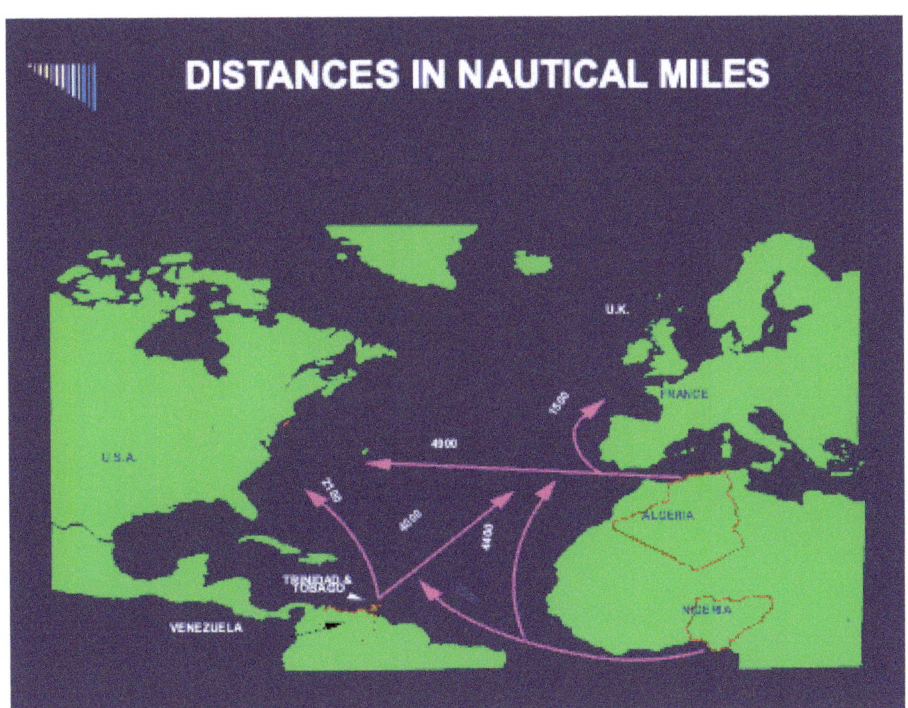

The Gas Market in the Northeastern United States.[7] Cabot's primary market area was New England, which accounted for over 90% of Cabot's sales. The ability of Cabot to market LNG from Trinidad and Tobago, and the price that could have been obtained for that gas, was driven by the supply and demand dynamics of theNew England gas market. New England's per capita gas consumption was second lowest among the regions of the United States, and Cabot had enjoyed significant strategic advantages to broaden its role in the various sectors:

- The company's receiving and regasification facilities were located downstream of the pipeline bottlenecks, which had served to limit gas distribution in New England.
- Cabot's business represented a major portion of the gas supply in New England, and Cabot could access virtually any market in North America.
- The terms of the Cabot sales contracts and Cabot's regulatory authorisation permitted it

---

[7] The information has been obtained from the 1995 Train 1 Financial Memorandum that was presented to the bankers.

significant price flexibility in the market, allowing it to compete against alternative supply.

- Unlike other gas suppliers and gas service providers, Cabot could have capitalised on the unique physical properties of LNG that allow for rapid delivery as a vapour or liquid.

- Cabot could respond quickly to customer needs, whereas alternative suppliers located at considerable distances from the Northeastern market had less operating flexibility to respond to rapidly changing market situations or to provide services on short notice.

New England Pipelines

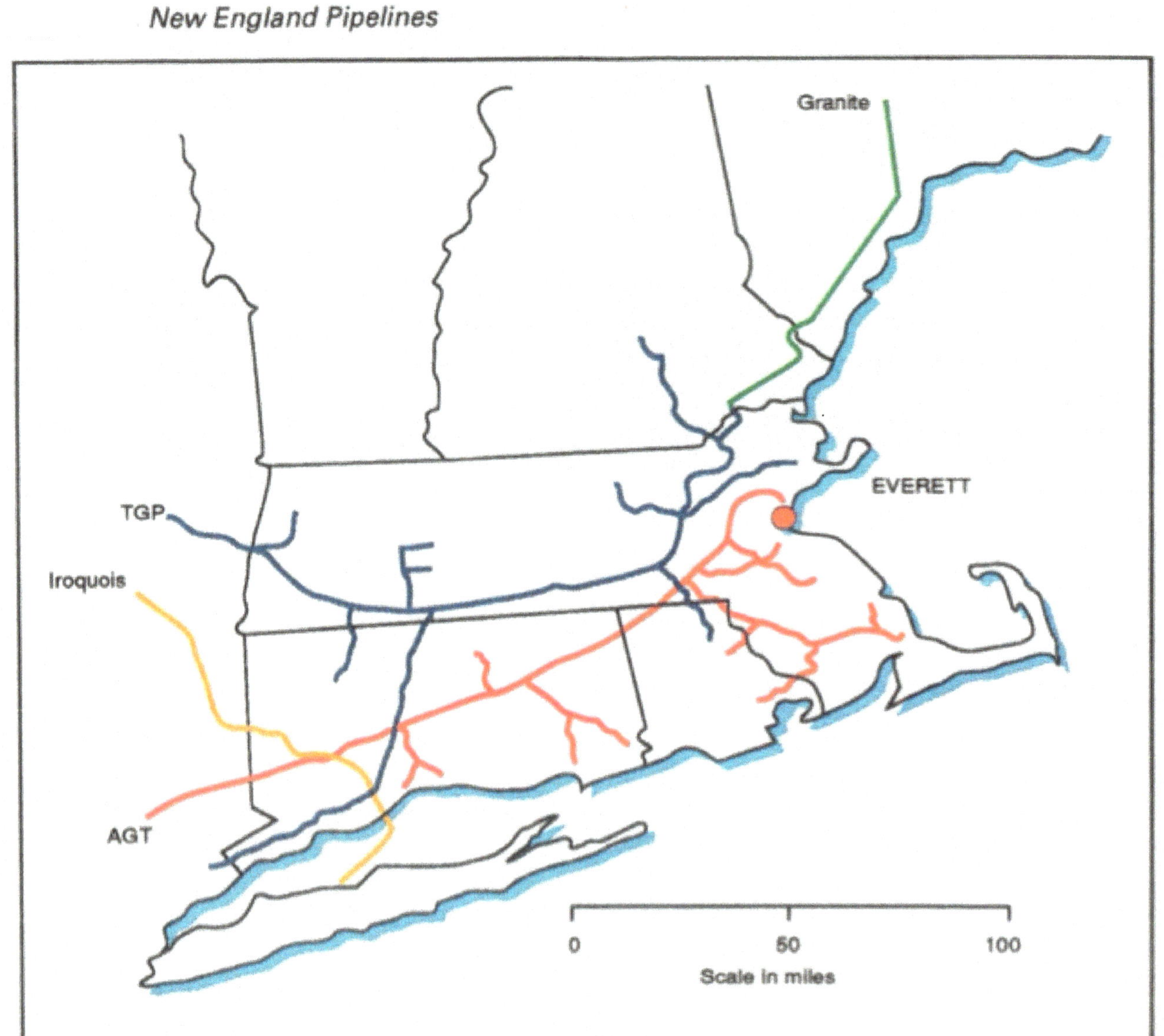

The Gas Market in Spain. Whereas natural gas demand had averaged 22% of primary energy demand in OECD countries, it had constituted just 6.4% of primary energy demand in Spain in

1993 and 6.8% in 1994. The Spanish Government's goal had been to increase the demand for natural gas to levels consistent with the OECD average. Estimates supplied by Enagas suggested that natural gas would represent 12% of primary energy demand by the year 2000. Spain was almost entirely dependent on imported gas supplies. In 1995, Algeria was expected to account for 49% of the total supply, and Libya a further 17%. In purchasing LNG from Trinidad, Enagas would have been diversifying and increasing its sources of gas supply. LNG from Trinidad was expected to constitute 8.5% of the total Spanish gas supply in 1999, with supplies from Algeria, under long-term contracts, constituting an estimated 61.5%.

Projected Natural Gas Supplies for the Period 1995 – 2005

Source	1995	1996	1997	1998	1999	2000
Algeria (LNG)	181	157	127	139	147	147
Algeria (pipeline)	0	32	167	190	226	238
Libya	62	60	60	60	60	60
Norway (pipeline)	56	49	62	79	79	79
Trinidad & Tobago (LNG)	0	0	0	0	26	57
Domestic	28	21	21	28	28	12
Other	43	81	69	45	14	53
Total	370	400	499	541	580	646
The forecasts were provided by Enagas; The quantities refer to MMBTUs						

LNG Terminals. In 1995, Enagas had three LNG receiving facilities, which were located in Barcelona, Huelva, and Cartagena. These terminals had an aggregate LNG storage capacity of 455,000 m³ and a regasification capacity of approximately 1.5 million m³ per hour. An additional terminal was planned for either the north coast at El Ferrol or the Portuguese coast, which would have created additional LNG storage capacity amounting to some 200,000 m³.

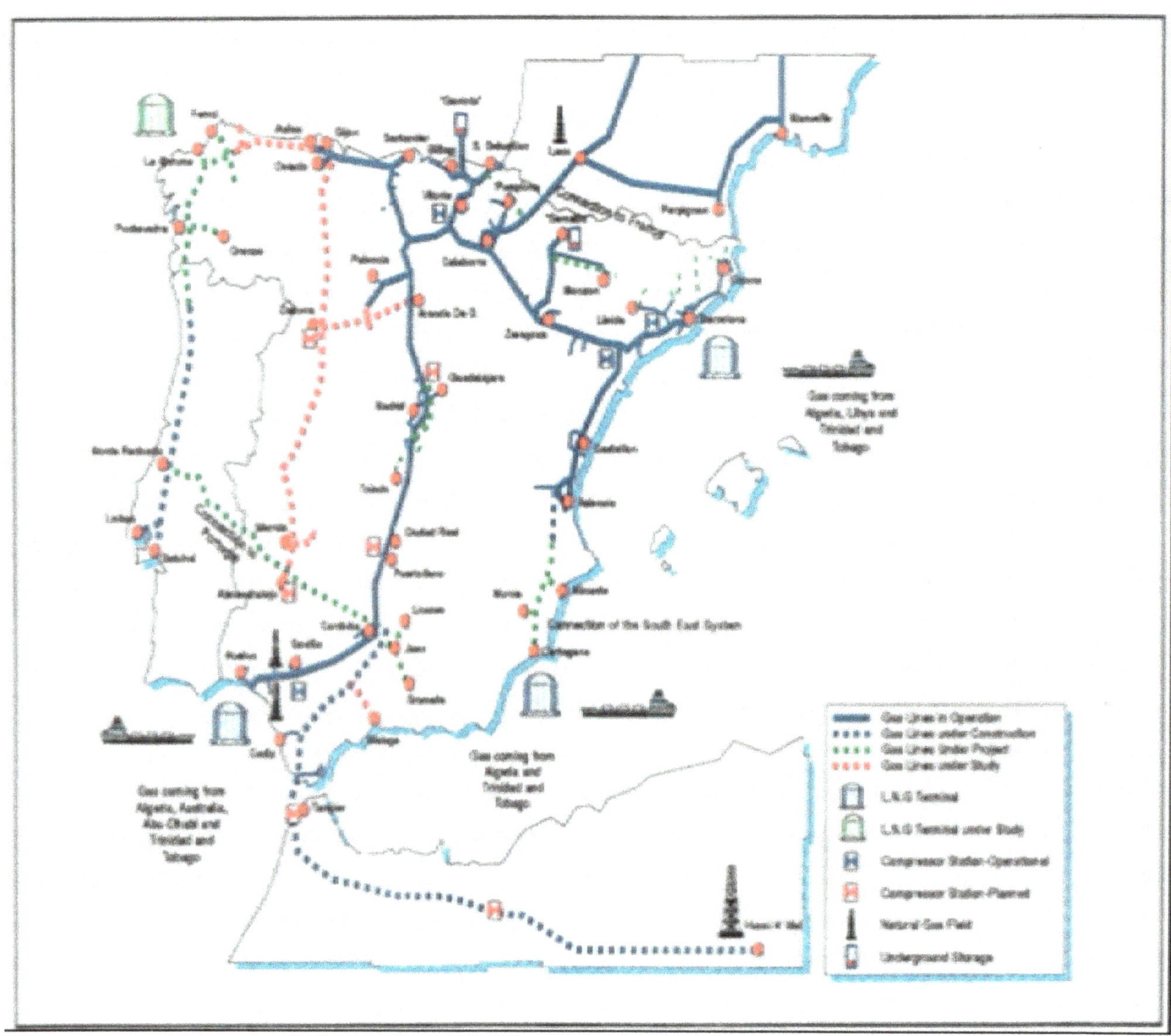

The LNG Plant Site. NGC's success in attracting natural gas-based industries meant that Point Lisas was approaching a capacity limit for huge, heat-generating industrial plants. Thus, the Government of Trinidad and Tobago chose a site for the LNG plant at Brighton, La Brea, on the southwest coast between San Fernando and Point Fortin. The area had a good natural harbour, which was thought to be a tremendous shipping advantage for the LNG project because of shipping constraints at the port of Point Lisas.

However, due to difficulties in site remediation at La Brea, on the 20th June 1996, pursuant to the 2nd supplemental Shareholders' Agreement, the sponsors agreed to change the site to Point Fortin. NGC was then discharged from any obligations for site remediation at La Brea.

Above: the La Brea deep water harbour. **Below**: the Point Fortin site, suitable for 6 LNG Trains.

Project Administration. When I took up my appointment as the LNG coordinator on 2nd March 1995, the project was administered by a Project Executive Committee ("PEC"), which was a committee that consisted of two representatives from each of the sponsors. The members of the PEC later became the representatives on the Board of Atlantic 1. The PEC had set up various project teams in order to implement the objectives identified in the MOU, and as the LNG coordinator at NGC, I was an ex officio member of all of the project teams. My role was to coordinate all of NGC's activities on the project, make commercial decisions in NGC's interests, and, when appropriate, advise NGC's representatives on the PEC (and later, NGC's representatives on the Atlantic 1 Board).

Atlantic 1 was formed on 20th July 1995 with an initial share capital of U.S.$30 million. In accordance with the Shareholders' Agreement, each Shareholder nominated a director and an alternate director to serve on the Board of Atlantic 1, and each director possessed votes in proportion to the number of shares held by the Shareholder that had nominated him. The first members of the Board of Atlantic 1 were:

Chairman   **Dr Kenneth Julien (NGC)**

Parent of Sponsor	Director	Alternate
Amoco	Mr. David Wight	Mr. William M. Janssen
BG	Mr. Martin Houston	Mr. Simon Bonini
Repsol	to be nominated*	to be nominated*
Cabot	Mr. Gordon Shearer	Mr. Geoff Hornby
NGC	Dr. Kenneth Julien	Mr. Malcolm Jones

* In the first supplemental shareholders' agreement, dated 30th November 1995, Antonio Garcia Mateo and Antonio Perez Collar were formally appointed as Repsol's director and alternate, respectively. Prior to Repsol becoming a shareholder, Cabot and BG transferred 15% and 5% interests, respectively, to Amoco who held the 20% interests on trust for Repsol. Thus, after Repsol became a shareholder, the shareholdings were adjusted to Amoco (34%); BG (26%); Repsol (20%); Cabot (10%); and NGC (10%).

ATLANTIC LNG COMPANY OF TRINIDAD AND TOBAGO

A Signing Ceremony

with our New Shareholder ·· **REPSOL International Finance B.V.**

Programme

2.00 p.m.	Prime Minister arrives
2.05 p.m.	Welcome: Mr. Martin Houston General Manager, British Gas Trinidad Ltd.
2.08 p.m.	Remarks: Mr. David Wight, President , Amoco Trinidad Oil Company
2.14 p.m .	Remarks: Mr. Gordon Shearer, President, Cabot LNG Corporation
2.20 p.m.	Remarks: Senor Antonio Garcia Mateo, Repsol International Finance B.V.
2.25 p.m.	Signing of Agreement
2.30 p.m.	Address: The Honourable Prime Minister of Trinidad and Tobago Mr. Basdeo Panday
2.40 p.m.	Vote of Thanks: Mr. Malcolm Jones, Managing Director National Gas Company of Trinidad and Tobago Limited

The President. Under the terms of a Letter Agreement dated 20th July 1995, the Shareholders agreed that Amoco would have the right to choose the President of Atlantic 1 for a period commencing from the date of the execution of the Letter Agreement to the first day of the first calendar month following the expiration of three years after Start-up of the Plant. Since, at the time, the Start-up of the plant was projected to be in 1998, it meant that Amoco would be able to choose the President until 2001 (effectively six years from 20th July 1995). Thereafter, the position of President had to be filled pursuant to the terms of the Shareholders' Agreement without regard to the provisions of the Letter Agreement.

Under the terms of the Shareholders' Agreement, the President, like the other officers of the company, had to be appointed by an affirmative vote of 66% vote of the directors. There was a restriction, however, that no person could be appointed as President if that person was, immediately prior to his/her term as President, employed by a shareholder (or its affiliate) whose director was serving as the Chairman. At various times during the course of 1995, the following first officers of Atlantic 1 were appointed:

Sponsor	Name	Office
Amoco	Mr. Gerry Peereboom	President
Amoco	Mr. David Jamieson	Vice President, Technical
Cabot	Ms. Mary Moore *	Vice President Law & Corporate Affairs
Repsol	Sr. Juan Varela	Vice President, Marketing
BG	Mr. Howard Candalet	Vice President, Operations
BG	Mr. Stephen Haynes	Vice President, Finance

* Ms. Mary Moore was also the secretary to the Board.

Letter Agreement: Gas Supply for Plant Expansions. In 1995, Atlantic 1 entered into a Letter Agreement with Amoco and British Gas concerning gas supply arrangements for LNG expansions.

My reservations about the Letter Agreement were communicated internally to NGC's representatives on the Board of Atlantic 1 and to Cabot.

"24th July 1995

Malcolm,

As matters stand, I am inclined to be cautious about future LNG trains because of the projected availability of reserves for domestic supply and the strategic need to ensure that the reserve base is sufficiently strong to support future petrochemical growth. However, in our leadership role, we should also encourage the development of the gas from the North Coast Marine Area, particularly as this could provide GORTT with revenue that it might not otherwise obtain.

A difficulty is that the Letter Agreement proposed by British Gas and Amoco does not limit the Expansion gas·to that from the North Coast. There could be a political backlash on the LNG Project if there is a public perception that reserves would be exported to the detriment of domestic needs."

"July 25, 1995

Mr. Geoff Hornby
President, Cabot Trinidad LNG Corporation
Algico Plaza
91 - 93 St. Vincent Street
PORT OF SPAIN
Fax. No: (809) 627 6976

Dear Mr. Hornby

Re: Gas Supply: Plant Expansions

I thought it appropriate to make a few observations in light of your proposed response to the Letter Agreement to be entered into between Amoco, British Gas, and Atlantic LNG Company of Trinidad and Tobago ("Atlantic").

I have focused my attention on the Shareholders' Agreement since that document reflects the existing agreement between the parties.

Voting Shareholders holding at least 51% of the Shares of the Voting Shareholders

Project Financing. The Plant, with a nominal capacity of 450 MMscfd, was designed to process natural gas supplied by Amoco from wells offshore the East Coast of Trinidad and produce approximately 3 million tonnes p.a. (or *circa* 400,000 MMBTU per day) of LNG. The Sponsors' key financing objective was to finance the construction of the Plant as a project financing transaction using a combination of limited recourse debt of $600 million and equity. In 1995, with the project cost assumptions, it meant a debt-to-equity ratio of 65:35. Equity was to be provided by the Shareholders in proportion to their respective ownership interests in Atlantic 1. Prior to the initial drawdown of bank loans, the Shareholders funded Project costs through equity contributions. During the drawdown period, the Project costs were funded by a combination of debt and equity. Financing was secured by a prior perfected lien on the assets of Atlantic 1 with no recourse to the Sponsors or Shareholders after Project Completion. The Project structure offered a comprehensive series of mitigating factors.

Project Economics. In the Base Case, the Loan Life Coverage Ratio (LLCR) was calculated using the all-in cost of debt as the discount rate. It was presented as of June 1999, when the Company expected to have fully drawn the $600 million of debt. The semi-annual Debt Service Coverage Ratio (DSCR) was calculated at December 1999, the date of the first repayment, and at June 2000, the second repayment date.

LLCR (Jun99)	DSCR (Dec 1999)	DSCR (Jun 2000)
2.01	1.24	1.63

LNG and NGL production were assumed to average 80% of plant capacity and 100% of plant capacity in the periods ending 31st December 1999 and 30th June 2000, respectively. Under various sensitivity analyses to the Base Case, debt could have been serviced on schedule.

The Technology Choice. In January 1993, a feasibility study was initiated by the Sponsors, and a FEED phase was subsequently launched in October 1994. At the outset, there were three bidders for the FEED phase:

(a) Chiyoda/Hudson and Kellogg/JGC, both of whom had chosen a process licensed by APCI, based on a mixed refrigerant process (the "APCI Process"); and

(b) Bechtel/Phillips, who had chosen an alternative, the "Optimized Cascade" process, developed by Phillips Petroleum.

The Chiyoda/Hudson consortium won the contract for the FEED.

In order to attract the most competitive bids for the main EPC Contract, the Sponsors decided to keep the options open for the two different liquefaction processes. Two teams (which were separated by a 'Chinese Wall') were retained to review the two processes independently.

Most of the recently-built LNG plants around the world had used the APCI process, which involved a propane pre-cooled mixed refrigerant process. The alternative technology, the Phillips "Cascade Process," used pure hydrocarbon refrigerants. The last plant that used the Cascade Process was a smaller plant with a capacity of 1.2 MTPA in 1969 in Kenai, Alaska. Thus, in a sense, the Cascade process was less 'known' technology, which had led to some concerns about whether the scale-up to the chosen design (3.0 MTPA) for the Project would be factors that militated against using the "Optimized" Cascade Process in Trinidad. And, even though the plant in Kenai had proven to be very reliable and had not missed a scheduled shipment since it began operations, there was some speculation that the APCI process would be the selected technology for the plant. In other words, the speculation was that the choice of contractor for the EPC contract would be either the Chiyoda/Hudson or Kellogg/JGC consortium.

On 8th March 1995, all three bidders were invited to tender ("ITT") for the main EPC contract, and each ITT made it clear that the contractors were expected to bear their own bidding costs. The Technical Team was excellent in their management of the bidding process, although one matter arose where, arguably, the Team was, perhaps, somewhat accommodating in their dealings with one of the bidders.

On 16th March 1995, Kellogg wrote the leader of the Technical Team advising:

> *"Since [our competitors] are presently being reimbursed for their specification effort, we believe this puts our group at a commercial and technical disadvantage. In order to provide a fair and competitive bidding process, we request reimbursement of a portion of our efforts at the time of the submittal of our proposal. Should we be the selected contractor and the project proceeds, we would credit this partial proposal reimbursement against our successful bid price. We believe this compensation should be $2,350,000."*

The Technical Team was willing to accede to Kellogg's request for various reasons, including a desire to ensure competition and better bid prices, but it was contrary to the terms of the ITT and risked protracted discussions with the other bidders. My advice to NGC's members of the PEC was as follows:

*"Since Chiyoda successfully won the bid for the FEED Contract (in preference to Kellogg/JGC) it has a competitive advantage in respect of the bid for the EPC Contract **if** the APCI process is selected. Also, if the APCI process is preferred and Kellogg/JGC do not bid, there would be only one bidder leaving the Parties no choice but to award the EPC Contract to Chiyoda/Hudson. There is, therefore, merit in preserving the option of a competing bid for the EPC Contract based on the APCI process in order to secure the best prices.*

Kellogg have suggested that they should be compensated for bidding costs to the tune of U.S.$2,350,000.00, which would be credited against their bid on the EPC Contract if they are successful. If the PEC does not accept these terms, there is a risk, which I find difficult to assess, that Kellogg would not pursue the invitation to Tender ("ITT"). A real issue, therefore, is whether the perceived advantages of having Kellogg/JGC as a competing bidder outweigh the disadvantages of having to write off the compensation sought in the event that the Kellogg/JGC bid is unsuccessful.

However, if one is prepared to offer Kellogg some compensation, a possible difficulty is how to justify the shift away from the stated position in the ITTs issued with release documents: it was apparently made clear that contractors were expected to bear their own bidding costs.

Chiyoda/Hudson might argue that they are bearing their own costs in respect of the bid for the EPC contract and that any advantage they have now is a result of winning the FEED contract (which was also open to Kellogg/JGC). Bechtel/Phillips might make the same point, although in their case, there were

The Award. I attended the opening of the bids on 18th July 1995 and the discussion after the subsequent evaluation of the commercial proposals in September 1995. The rate of return to shareholders for the project was very sensitive to Plant costs. Thus, the EPC Contract was awarded to the lowest bidder, Bechtel, whose lumpsum price proposal was less than the next bidder's proposal by more than U.S.$ 100,000,000. The Technical team oversaw the work of Bechtel until the Plant passed all performance tests at completion. Train 1 first produced LNG on 11th April 1999, and the first cargo left Trinidad for delivery to Cabot in Boston on 1st May 1999.

Atlantic LNG Train 1 Commercial Negotiations. The key commercial agreements were (a) the LNG Offtake Agreements between Atlantic 1 and Cabot and Atlantic 1 and Enagas; (b) the Gas Supply Agreement between Atlantic 1 and Amoco; (c) the Project Agreement between Atlantic 1 and the Government; and (d) the commercial agreements with NGC.

(a)   The Cabot LNG Offtake Negotiations. The Atlantic 1 team (the "marketing team") that negotiated the LNG Offtake Agreements with Cabot and Enagas consisted of representatives from British Gas, Amoco, and NGC. The project was about to collapse, and a top-level delegation of the proposed investors met on Easter weekend in 1995. NGC was represented by Mr. Malcolm Jones, who was accompanied by Mr. Manning, the then Prime Minister. As a result of the meeting, the parties scheduled negotiations between the marketing team and Cabot for mid-May 1995 at Amoco's Westlake offices in Houston, Texas.

The Cabot negotiations was, for me, a baptism by fire. Before leaving for the meeting in May, Malcolm Jones (in his typically laconic style) had said to me:

> *"Do not bother to come back if the negotiations break down. Go straight back to England. We have nothing here for you in Trinidad."*

The Issues. Much earlier, towards the end of March 1995, I had received a copy of "General Principles (3/24/95)" for the Cabot Corporation Guarantee and the Cabot LNG Corporation commitments for the Atlantic LNG sales contract. Under these "General Principles," it was said that "Cabot LNG will guarantee" that adequate Shipping Capacity will be provided to be capable of lifting its contracted ACQ during the term of the LNG Sales Contract. My first reaction was to object to the entity purporting to provide the "guarantee" for adequate Shipping Capacity. In my view, the guarantee should have been given by the company in whom legal title to the ship was vested (or the person who owned that company).

Accordingly, I had formally advised NGC's PEC representatives that –

> *".... if the ships in question were owned by Cabot LNG Shipping Corporation (which was certainly the case with the "Gamma"), it would have been more appropriate for the guarantee to be given by the parent company - Cabot Corporation - or at the very least, the legal owner of the ship(s) - Cabot LNG Shipping Corporation. ..... ."*

My objections about the Cabot guarantee were superficial, as borne out by the note of a meeting on 28th March 1995 among project representatives from BG and Amoco, and Cabot:

> *"Cabot Negotiation Meeting - March 28, 1995*
>
> *Boston, MA - **Project Reps**: Alan Stott, R. E. Sloan, R. K. Stehn. **Cabot**: Gordon Shearer, Paul Saba, Ted Gehrig, Bob DeAngelis (& Kennett F. Burnes for approximately 10 minutes).*
>
> *Kenneth Burnes made the introductory comments and stated forcefully that we were hearing from Cabot Corporation today, and that included Mr. Bodman, John Cabot, and himself, and he wanted no misunderstanding about that. Further, he said that Cabot would respond to all points of our Cabot Corporation Guarantee proposal dated March 24, 1995, and that the response would be negative.*

He said that this concluded his comments and that he would be available the rest of the day Wednesday, and that Mr. Bodman would be in the office on Wednesday should we wish to discuss any issues with either of them. He then turned the meeting over to Gordon, and he left the meeting.

Gordon's opening comments were as follows:

He restated that the response was a Cabot Corporation response and that it contained three elements, two were buyer issues, and one dealt with the economic environment of the plant.

- *The project price split proposal of 19.9% to Cabot (80.1% to the Plant) was not acceptable to Cabot, and they were unwilling to discuss it any further.*

- *On the guarantee, he said t h a t as Ken Burnes has maintained, Cabot Corporation w i l l n o t give a guarantee in the form of our proposal (Attachment 1).*

- *Based on their estimates of an average FOB price to the plant and 39-40% going to the wellhead, they will not participate in the plant as an investor.*

Gordon insisted that this doesn't mean that a deal cannot be made. From Cabot's point of view, there is:

- *Very little room to move from their December 1994 price split offer.*

- *Cabot is willing to put at risk all the assets of the LNG business together with operating the assets and will back the upgrading of the assets.*

- *May consider a cash supplement (i.e., might be willing to put up some cash as well as the LNG assets).*

- *If the economic structure of the plant is provided, appropriate economics Cabot is interested in investing.*

- *Ken Burnes had a strong impression from his meeting with Amoco in Houston on March 20th that Amoco was very concerned about ceding control of the LNG marketing to Cabot.*

Gordon then presented other alternatives, such as:

- *Project or ?? (sic)[8] buying 50% of Cabot LNG and Cabot either using or not using proceeds to buy into the plant or*
- *Contribute all LNG assets to the project and, in return, obtain an interest in the project or*
- *Out-and-out purchase*
- *Cabot can be a buyer of gas only with no interest in the plant – Cabot's primary interest is to acquire LNG at an attractive price."*

Later I discussed the note of the meeting with an Amoco representative on the Marketing team who, to my surprise, was more nonchalant than I had anticipated. His reaction to my concerns was that Cabot was "playing games," and there might be a corporate strategy to "maintain an interest in LNG whilst limiting the risk exposure." I asked him how he viewed the various alternatives suggested by Gordon Shearer, in particular, the possible acquisition of interests in Cabot. He told me that there was strong support for this in Amoco, but they would have to do some "soul searching." When pressed, he indicated that not everyone was in favour, but on balance, the signs were favourable. Subsequent events below demonstrated that Cabot Corporation, at that time, were seriously contemplating its departure from the LNG business:

- During 1995, Cabot reduced its stake in the project from 25% to 10%;
- Cabot LNG separated from Cabot Corporation in 1999; and
- Cabot LNG eventually sold its 10% interest in Atlantic LNG to Tractebel.

And, as the project structures of subsequent LNG expansions have demonstrated, at that time, Amoco had already had a strong interest in vertical integration into the LNG marketing end of the business. In any event, after the Cabot meeting in March with Amoco, the issues further escalated, which led to the Easter Monday meeting that included Malcolm Jones and Mr. Manning.

The Cabot Negotiations. Prior to the negotiations in May, I wrote Malcolm Jones and expressed to him my views on possible solutions to Cabot's proposals for the Cabot Corporation guarantee and the netback split to Atlantic 1. In my analysis, after the risks of other options

---

[8] There was an obvious typing error in the document. Was the word "team"?

were evaluated, it seemed commercially prudent to make concessions to Cabot on both issues.

"Cabot Corporation Guarantee

The investors in the Plant have sought a guarantee from Cabot Corporation of sums relating to contractual quantities. In response, Cabot, apparently, has been prepared to put up the value of its LNG assets, including the ship, as a guarantee. This guarantee, if accepted, would exclude from Cabot any liability for sums over and above the value of those assets arising from Cabot's failure to take LNG due to, for example, a depressed New England market. Cabot has also undertaken to bring the Everett terminal up to the capacity to take the plant output and refurbish the ship.

Notably, everyone is prepared at the outset to use the New England market to determine the Netback to the Plant: it is that market (which either exists or does not exist) that is the whole thrust behind the Cabot LNG sales contract. But the acquisition of the Cabot LNG assets and the ship (if that is an option) as an alternative to Cabot's participation necessarily carries risks associated with the use of those assets. Also, any such acquisition does not change the position with respect to the possibility of a failure of the New England market. Indeed, the Project would still be then accepting the risk of market failure as well as (without input from Cabot) an additional risk that the expertise on marketing and maintaining the terminals would depart if Cabot did not participate. But perhaps the latter element could be minimised to some extent by some degree of Board control by buying into Cabot LNG. Accordingly, unless Citibank is opposed, I believe that it would be appropriate for NGC to be prepared to support an even greater move towards Cabot's position on the guarantee. The key would be to attempt to identify the extent of the Investors' exposure as a result of the proposed guarantee and then take steps to minimise the resultant risks.

As to the quantum of the guarantee, if the New England market failed to the point where Cabot could not recover its costs and lost money, Cabot could, in the first instance, continue to take (and pay for) all the contracted volumes of LNG. But it is not inconceivable that at some stage, it would become commercially impossible for Cabot to take all, or any, of those contracted quantities. In such a case, Atlantic LNG would be obliged to mitigate its losses by, for example, taking steps to sell any relevant quantities in other markets, if any. In other words, the measure of damages would then be determined by reference to mitigation by Atlantic LNG of its losses. Arguably, in those circumstances, the need for a guarantee from Cabot Corporation in respect of the entire sum due under the take or pay provisions would be an extreme position.

Price

*On price, it is clear that the Project has accepted an offer from Enagas (40 % Plant output) that is less than that from Cabot (60%). Also, other European buyers are apparently **not** prepared to offer a higher price than Enagas. Presumably, the position would improve if Puerto Rico came onstream in due course (assuming a higher price), but even then, it is not entirely clear that Puerto Rico would be able to take the quantities displaced by Cabot. In essence, therefore, Puerto Rico would help, but it might not solve the problem in its entirety, particularly if the time frame (1988) for project completion is crucial. Accordingly, if Cabot were to withdraw as a Buyer, the likelihood is that the Project would have to consider an Enagas selling price for the whole output as the alternative. The economics, clearly, would not be then more favourable than a move now towards Cabot on price and obtaining a resulting selling price higher than that for Enagas or other European buyers."*

Later, I provided Malcolm Jones with a written status report of the negotiations at Amoco's offices in Westlake, Houston, and noted that, after two weeks, there was closure on most of the contractual issues. However, the most important outstanding issues had still related to the corporate guarantee from Cabot Corporation. During the 2nd week, another matter, the escalator for the "marketout" sales price of gas in New England, had also surfaced.

*"**Market out**. Amoco produced graphs that suggested that below an LNG sales price of $2.30 in the U.S., there would be a negative impact on Cabot's net operating cash flow. The predictions were that the Plant would then have even greater losses in that price range. Accordingly, the Marketing Team proposed to Cabot (who accepted) a market-out trigger mechanism at gas prices of $2.00, $2.07, and $2.15 that would apply for the years 1998, 1999, and 2000, respectively. Trigger prices in respect of subsequent years, it was agreed, would be tied to an escalator. It was therefore left to the project team to ensure that any chosen escalator would not generate a trigger price that would permit Cabot to 'market-out' whilst making profits.*

***Corporate Guarantee**. On 3rd May 1995, Cabot had proposed a draft guarantee, the terms of which would have enabled Cabot Corporation, among other things, to "buy-out" of the guarantee for $140 million (inclusive of the value of the ship) if Atlantic LNG were to claim against the guarantee. My immediate reaction to that proposal was that the option had limited Cabot's total exposure to $140 million (and not $200 million as agreed earlier). However, after discussion within the team, I appreciated that $140 million upfront could be equivalent to $200 million achieved after prolonged arbitration.*

My main concern, therefore was to ensure that the Cabot Corporation guarantee would not let Cabot LNG off the hook if it decided, unilaterally, to fail to take LNG because of better profits with another Seller. This point was relevant whether one was considering an agreed cumulative cap of $200 million or an alternative immediate payment (including the value of the ship) of $140 million. In either scenario, Cabot LNG could simply fail to take LNG and leave it to Atlantic LNG to pursue claims against Cabot Corporation until either the $200 Million guarantee limit had been reached or Cabot Corporation had exercised the buyout option. It was not inconceivable that the Plant would then be exposed, for example, if it were obliged to mitigate by selling to Europe (with lower prices)."

I had indicated to the marketing team that if the buyout option were to be rejected:

Under that scenario, Amoco had proposed that one option would be: "*Keep the LNG sales contract alive and, under arbitration, seek an award requiring Cabot to specifically perform its obligations to take LNG.*" I had serious doubts that the remedy of specific performance would be available to Atlantic 1, particularly in circumstances where the contracting parties would be deemed to have clearly contemplated damages (and even so, limited damages) as the remedy for breach of contract.

During the negotiations, Cabot (Paul Saba) had said, very dramatically, that the negotiations would be terminated unless Atlantic 1 accepted that the corporate guarantee had provided a global limit in respect of the obligations of the Cabot companies. Eventually, Cabot's position was accepted, although I suggested to the team that the obligations of Cabot LNG should be discharged under the terms of the Cabot Corporation guarantee in very limited circumstances, for example, only where the parties would justifiably consider termination. That approach, I had opined, would be "*consistent with the decision to lock into the New England market and allow Cabot a contractual 'economic out' in narrow specific instances relating to non-profitability.*"

Relationships. I found Gordon to be an excellent and clear-thinking negotiator and Paul, a lawyer's lawyer, who strove for perfection. Indeed, Paul's mantra was: "Always keep reading the document." In tandem, Gordon and Paul were a lethal combination for any opposition. Over the course of the 2-week period, I took the opportunity to convey NGC's views privately to Gordon and Paul that the relatively low cap of the $200 Million corporate guarantee should not be used as a mechanism to buy out the long-term take or pay contract.

Despite the tense atmosphere, there were humourous occasions. For example, as a follow-up one evening, I reminded Paul that earlier, during the negotiations, when we were close to obtaining a concession from Gordon (in response to a proposal from Amoco), he (Paul) had cleverly interrupted Gordon on the pretext that he (Paul) did not understand Amoco's proposal. Paul then said to me: "*So you picked that up. Gordon talks too much.*" On another occasion, knowing that Gordon had Scottish roots, I told him that my favourite drink was scotch and coconut water, to which he replied: "*René, how can you ruin those two excellent drinks by mixing them?"*

The main Contractual Provisions Agreed with Cabot

- Destination Flexibility.[9] Flexibility for Cabot and Enagas to exchange cargoes insofar as such exchanges did not undermine the very essence of the contract, i.e., sales of LNG into the higher value New England Market. Thus, there were minimum volumes of LNG that Cabot must have made reasonable efforts to deliver through its Everett terminal during the Summer and Winter periods.

- Netback Fraction ("NBF"). The FOB price payable for the LNG was structured as a "netback" arrangement under which the FOB price was directly related to the tailgate price received by Cabot for LNG marketed through the Everett terminal. The tailgate price was calculated on a semi-annual basis (Winter and Summer) which the NBF was applied to calculate the FOB price to be paid to Atlantic 1.

Tailgate Price	Netback Fraction
Less than or equal to $2.90 per MMBTU	0.76
Between $2.90 and $3.10 per MMBTU	Linearly Between 0.76 and 0.74
Equal or more than $3.10 per MMBTU	0.74

The NBF could have been increased by up to 0.0135 if Cabot received LNG deliveries of up to 25.2 million MMBTU from other LNG suppliers. Sales of LNG to Cabot affiliates, which did not market or resell LNG, were subject to a separate review by Atlantic 1 and, if not approved, were excluded from the price calculation.

- Market Out. The contract could be terminated due to persisting and severely adverse economic impacts to Atlantic 1 or to Cabot caused by, for example, a badly deteriorated market, new taxes, or unforeseen capital expenditures mandated by new regulations.

---

[9] Following an arbitration decision in 2008, for the calendar quarters when the majority of Enagas cargoes were shipped to the Everett terminal, the price of LNG for all contracted volumes, even those destined for Spain, was based on the Boston City Gate price. If the majority of cargoes were shipped to Spain, all LNG for that quarter was priced according to the Spanish pricing formula.

- <u>The Corporate Guarantee</u>. Cabot Corporation would execute a guarantee in favour of Atlantic 1 for Cabot's monetary obligations under the Cabot contract, and the maximum aggregate liability of Cabot Corporation and Cabot was $200 million.

- <u>Condition Precedent</u>. Under Article 24 of the Contract:

*"By 30 September 1995, [Atlantic 1] must have executed a Gas Supply Contract with [Amoco], in form and substance satisfactory to Amoco and [Atlantic1], in the sole discretion of each of them, which Gas Supply Contract will be provided to [Cabot] for its review and approval in connection with its termination right as **provided** in Article 24.3(b)."*

Under Article 24.3(b), Cabot was entitled to review the Gas Supply Contract with Amoco to determine whether, in its reasonable opinion the Contract adequately addressed Cabot's commercially reasonable concerns as an LNG Buyer. Thus, through Cabot, Atlantic 1 had held significant leverage over Amoco when BG, Cabot, and NGC negotiated the feedstock gas supply contract for Atlantic 1. The Cabot LNG contract was executed on 27th July 1995.

IN WITNESS WHEREOF, each of the Parties has caused this Contract to be executed by its duly authorized officer as of the date first written above.

SELLER:
ATLANTIC LNG COMPANY OF
TRINIDAD AND TOBAGO

By _______________________

BUYER:
CABOT LNG CORPORATION

By _______________________

(b) The Enagas Negotiations. The discussions with Enagas in Spain were relatively quickly concluded after the Cabot negotiations. Some of the provisions of the Enagas Contract (as distinct from those in common with the Cabot contract) are summarized as follows:

- *Quantities*. Enagas' ACQ of 56.8 million MMBTU was also divided into seasonal volumes.

- *Price*. The contract price applicable for each calendar quarter was calculated on the first day per calendar quarter in U.S. Dollars per MMBTU in accordance with a formula based upon the monthly average prices of fuels published in Platt's Oilgram Price Report.

- *Contract Price Reopener*. The price formula was subject to reopener provisions if either Atlantic 1 or Enagas believed that the formula no longer reflected the value of natural gas in Enagas' end-user market. The first reopener could not be requested until 12 months after the date of First Commercial Supply, and subsequent reopeners were required to be spaced at least three years apart.

- *Liabilities*. If Atlantic 1 failed to deliver the ACQ to Enagas, Atlantic 1 became liable to Enagas in an amount equal to 25% of the Contract Price multiplied by the shortfall quantities. Enagas' liability was limited to its "take-or-pay" obligation.

The Enagas contract was executed on 26th July 1995.

IN WITNESS WHEREOF, each of the Parties has caused this Contract to be executed by its duly authorized officer as of the date first written above.

SELLER:
Atlantic LNG Company
of Trinidad and Tobago

By _________________________

BUYER:
Enagas, S.A.

By _________________________

O:\PBR\DOCUMENT\ENA-LNG9.DOC
26 JULY 1995

The Gas Supply Negotiations. At the 1st meeting of the gas supply teams, on Monday, 6th March, BG, as a potential gas supplier, sat with Amoco. However, at the 2nd and subsequent meetings, BG joined the team that negotiated the feedstock agreement on behalf of Atlantic 1.

The gas supply negotiations should be viewed in the context of timing constraints and conflicts:

(a) Timing. The date agreed for concluding negotiations, 30th September 1995, was important because it would have satisfied a condition precedent in the Cabot LNG offtake agreement and facilitated financing and project completion in 1998.

(b) Conflicts. There were many conflicts of interest within the negotiating team (BG, Cabot, and NGC), which were not manifested within the marketing team (BG, Amoco, and NGC) during the LNG offtake negotiations with Cabot and Enagas:

    i.   BG, as an investor in the Train 1 Plant, ostensibly, had aimed to secure the lowest gas price on behalf of Atlantic 1 (and thereby increase the return on its investment in the LNG Plant). However, BG's perspectives were also influenced as a gas supply competitor to Amoco in Trinidad.

    ii.   Cabot, like BG, had aimed to secure the lowest gas price on behalf of Atlantic 1, but by virtue of the condition precedent in its LNG contract with Atlantic 1, Cabot had kept its commercial interests alive to negotiate, if appropriate, better terms for that LNG offtake agreement.

    iii.   NGC was, perhaps, the most conflicted party, to the point of being schizophrenic. NGC's strategy was to achieve the highest well-head gas price - an Amoco objective on the other side of the table - provided that the price would facilitate a reasonable (but not the highest) rate of return for investors in the Plant. This strategy is borne out by my internal memo on 12th April 1995:

"It would be logical to assume that NGC (on behalf of GORTT) would bear in mind GORTT's view to have the profitability at the wellhead (as opposed to the Plant) because of the benefit of taxation at the wellhead and the tax holiday at the Plant.

Accordingly, NGC's position should necessarily be a compromise: the maximum selling price of gas to the plant that would still permit the Plant to be economical. We, therefore, have the difficult objective of securing the highest selling price of gas in circumstances where [NGC, BG, and Cabot] should, in principle, seek to have maximum profitability for the Plant (and therefore, the cheapest cost of gas). The difficulty is compounded because NGC is ostensibly arguing on behalf of the Plant during these gas supply negotiations."

By 1st June 1995, tensions within the team began to surface. BG ($0.85 per MMBTU) and Cabot ($0.65 per MMBTU) proposed gas prices that would increase their respective returns on their investment in the Plant. In contrast, NGC proposed that the gas price should be at least $1.00 per MMBTU, reasoning that it was important for the sponsors to ensure that Trinidad and Tobago derived the optimum benefits from the Project through wellhead revenue for the Government and the strategic availability of additional sources of ethane and NGLs.

On Tuesday, 22nd August, Amoco proposed a 45% netback split to the wellhead (a reduction from 47%) as a counter to the 41% that had been proposed by BG and Cabot. At the time, Plant economics were still uncertain. Thus, from NGC's standpoint, a direct relationship between the netback split and the capital costs of the Plant meant that there was an indirect transfer of risk to the Government when the capital costs of the Plant were unknown. Eventually, Amoco tabled a proposal based only upon BTUs supplied, which was the genesis of the pricing agreement.

The Atlantic LNG Pipeline. On 15th September, the impact of the capital costs of the Plant (which had been simmering below the surface) eventually came to a boil. Cabot and British Gas were finally prepared to accept a purchase price of gas from Amoco based upon the BTUs supplied. However, they still argued that based upon the netback to the Plant in respect of LNG sales, the subsequent netback to Amoco, in respect of the sale of the feed gas, would not leave the shareholders in the Plant with sufficient revenue to cover relevant costs and still provide them with a reasonable return because of the projected capital costs of the investment in the Plant.

Thus, Cabot and British Gas contended that in view of Amoco's greater upstream profits arising from the netback split, Amoco should pay a tariff (or some form of capital contribution) to support Atlantic 1's investment in the onshore pipeline. Amoco countered by proposing that it would assume ownership of the onshore pipeline and so eliminate Atlantic 1's exposure to the then estimated U.S.$ 62 million capital cost for the onshore pipeline. This decrease in the capital costs in Atlantic 1 would have increased the returns to all investors (including BG and Cabot).

Under the Shareholders' Agreement, Atlantic 1 had been formed "*...to design, construct, own and operate the Plant **and the pipeline from Abyssinia to the Site**.....*" [Emphasis added]. For different strategic reasons, Amoco's proposal to own the onshore pipeline was a bitter pill for both NGC and BG to swallow. On 18th September, BG wrote Dr Julien (in his capacity as chairman of Atlantic 1) expressing its strong objection to Amoco's proposal noting, among other things, that –

(a) *Atlantic 1 should be able to accept gas at Abyssinia to encourage free and fair competition for future expansion gas requirements.*

(b) *It was a far less complicated arrangement than having the line owned by a single gas producer who intends to use it for transportation to customers other than Atlantic 1.*

(c) *Amoco would be using its large upstream profits generated by Atlantic 1's generous netback arrangement to gain control over Atlantic 1's future supply options.*

(d) *Amoco's positive strategic value created negative value for Atlantic 1 in the form of loss of control/free access and political/legal risk.*

(e) *Atlantic 1 should be able to freely negotiate with any party for the supply of gas without requiring consent and payments to Amoco - a conflicted party.*

However, the 30th September deadline was rapidly approaching, and British Gas capitulated on the 21st September. NGC, to avoid a risk to its transportation monopoly, responded:

However, NGC, with 10% of the votes in the Atlantic LNG forum, did not prevail.

Dedication of Gas Reserves.[10] During the development of Train 1, Trinidad and Tobago's commercial gas reserves had been concentrated in the ECMA. The breakdown of proved, probable, and possible gas reserves in the two offshore regions is shown in the table below.[11] At the time, the NCMA was not in production. The Fields that were dedicated to the Project, namely, East Mayaro (EM) and South South East Galeota (SSEG), contained a total of approximately 4.6 Tcf (which was approximately 26% of the total reserve base in Trinidad and Tobago). The use of proven gas reserves for Train 1 (approximately 4.6 Tcf out of 17.6 Tcf) did

---

[10] The information was taken from the Financial Memorandum that was presented to the Lenders.
[11] *Ibid.*

not raise any concerns for NGC about the long-term security of supply.

*Gas Reserves** (Bscf)

	Operator	Proved	Probable	Possible	Total (Bscf)
FIELDS DEDICATED TO THE PROJECT[1]					
East Mayaro	*Amoco* [2,3]	1462	184	573	2219
South SEG	*Amoco* [2,3]	1635	424	300	2359
EAST COAST MARINE AREA					
Cassia, Teak, Samaan, Poui, Flamboyant, Immortelle, Banyan	*Amoco* [2,3]	1935	434	1070	3439
Dolphin	*British Gas/ Texaco*[4]	2130	490	510	3130
Kiskadee	*Enron* [5]	750*	0	0	750
Other Fields	*Amoco* [2,3]	800	270	2610	3680
NORTH COAST MARINE AREA					
Chaconia, Hibiscus	*British Gas* [4]	2000	0	0	2000
Total T & T		10712	1802	5063	17577

Gas Supply Security. There were many factors that contributed to Atlantic 1's gas supply security.

1. *Remaining Reserves*. The table below [12] shows the forecast of annual gas production from SSEG and EM Fields. The gas reserves remaining in the ground at the final maturity of the Project debt were substantial, and 4.8% of the proved reserves remained at the end of the 20-year contract.

Forecast of Annual Production from the Dedicated Fields

---

[12] Ibid.

Year	Total Annual Production Requirement (Bscf)	S.SEG Production Rate (MMscfd)	S.SEG Annual Production (Bscf)	EM Production Rate (MMscfd)	EM Annual Production (Bscf)	TOTAL Annual Production (Bscf)	Proved Reserves Remaining (%)
0	0.0	456	0.0	0	0.0	0.0	100.0
1	158.0	447	162.6	0	0.0	162.6	94.7
2	158.0	417	157.9	0	0.0	157.9	89.7
3	158.0	342	137.3	112	21.6	158.9	84.5
4	158.0	311	113.4	162	45.2	158.5	79.4
5	158.0	270	110.0	160	51.7	161.8	74.2
6	158.0	227	92.1	228	67.3	159.4	69.0
7	158.0	268	88.5	196	71.4	159.9	63.9
8	158.0	210	81.9	241	80.0	161.8	58.6
9	158.0	285	82.6	182	74.0	156.6	53.6
10	158.0	261	96.8	165	61.7	158.5	48.5
11	158.0	224	96.0	196	65.9	161.9	43.2
12	158.0	151	67.0	292	92.1	159.1	38.1
13	158.0	156	54.3	282	103.5	157.8	33.0
14	158.0	181	59.4	293	98.6	158.0	27.9
15	158.0	163	57.4	312	101.8	159.2	22.8
16	158.0	120	50.9	308	106.0	156.9	17.7
17	136.0	83	36.4	248	100.7	137.1	13.3
18	104.0	58	25.3	192	79.1	104.4	9.9
19	82.0	117	23.9	136	58.8	82.7	7.2
20	75.0	92	38.1	77	37.0	75.1	4.8
Totals	2,925.0		1,631.8	1,316.5	2,948.3	1,631.8	

2. *The Development Plan.* Two significant features of the development plan ("Exhibit E" to the gas supply contract) were:

    i. The initial development of SSEG would allow deliveries of over 500 MMscfd for the early years of the Project.

    ii. After the LNG plant became operational, Amoco would maintain deliverability in excess of its contractual requirements as a cushion against any unforeseen operational problems.

3. *Reserves Audit.* Gaffney Cline had assessed the quantities of reserves within the dedicated areas and had concluded that the proved reserves were sufficient to allow Amoco to satisfy its minimum obligations during the primary term of the Contract.

4. *Force Majeure*. Under the *Force Majeure* provisions:

- *"... loss or failure of the Gas reservoirs included in the Reserves and the deliverability associated therewith due to natural depletion or the absence of economically recoverable Gas (other than such an absence which itself is the result of Force Majeure) shall not be considered to be an event of Force Majeure."*

- *".. a failure of any Reserves Audit Report or Expert determination to accurately assess the quantity of Gas included in all or any portion of the Reserves or the deliverability associated therewith shall not operate as an excuse of any obligation to deliver Gas hereunder."*

The gas supply contract was executed by David Wight and Kenneth Julien on 4th October 1995.

(c)The Government Negotiations. In October 1995, Prime Minister Patrick Manning called a General Election, which was held in November 1995. The ruling PNM and the UNC opposition both won 17 seats. The UNC was able to form a coalition with the two-seat National Alliance for Reconstruction, allowing the UNC leader, Basdeo Panday, to become the Prime Minister. Ken Julien immediately resigned as NGC's chairman. Mr. Finbar Gangar became a senator and the Minister of Energy and Energy Industries.

In February 1996, a new Board was appointed for NGC, which was chaired by Mr. Steve Ferguson but did not include Malcolm Jones, the former managing director. Thus, NGC, a key shareholder, was then without the two leaders who had pioneered the project - Ken Julien and Malcolm Jones. Negotiations with proposed Lenders in respect of project financing were outstanding. And, under the terms of the Shareholders' Agreement, the Government's approval and various fiscal incentives were required before the shareholders would incur additional expenditure by awarding the EPC contract. The former NGC Board had believed that the Government's support for the project would send a positive signal to the international financial community that Trinidad and Tobago was a premier location for investment. Would the new Administration support the Project?

One of the first tasks of the new Board was to review the LNG Project, and as the LNG Coordinator, with single-point executive responsibility, I was called to many Board meetings over a period of around four weeks to brief the new members of the Board. After the Project was reviewed in its entirety, NGC's Board concluded that there were three issues to be resolved as pre-conditions for its recommendations to the Government for its approval of the Project:

(a) The type of fiscal and other incentives that would be appropriate for Atlantic 1;

(b) The ownership of the onshore natural gas pipeline to supply Atlantic 1; and

(c) NGC's future requirements for natural gas to its customers in the 'domestic market'.

Ultimately, NGC's Board, under the auspices of the Ministry of Energy, held daily negotiations at the Hilton Hotel with the other project sponsors. These negotiations, which were chaired by Steve Ferguson, addressed the three pre-conditions for Project approval and culminated in 3 Agreements:

(a) The LNG Project Agreement to be entered into between the Government and Atlantic 1;

(b) Principles of Agreement relating to operation and ownership of the pipeline that would supply natural gas from Abyssinia on the East Coast of Trinidad to the LNG plant at Point Fortin (the "Pipeline Principles"); and

(c) Principles of Agreement for amending a 1991 gas sales contract between NGC and Amoco (the "Gas Supply Principles").

Rupert Mends, then the Permanent Secretary in the Ministry of Energy, and I, assisted John Andrews in the negotiations for the LNG Project Agreement between the Government and Atlantic 1. The negotiations in respect of the Pipeline Principles and the Gas Supply Principles were led by two other NGC senior executives: Clarence Harnanan and Frank Look Kin, respectively.

The LNG Project Agreement. Under the terms of the LNG Project Agreement, the Government granted concessions to Atlantic 1 that included the following:

(a) A ten-year tax holiday under the Fiscal Incentives Act (FIA) with respect to –

    i. Corporation tax;

    ii. Withholding tax on dividends and other distributions;

    iii Customs duties on major capital goods, excluding spare and replacement parts; and

    iv. Value-added tax on imported goods.

(b) Relief from interest withholding tax over the ten-year tax holiday period.

(c) An adjustment of the depreciation rate on plant and machinery.

(d) Relief from municipal property tax.

(e) With respect to the site for the Plant:

    (i) A purchase price of U.S.$7.5 Million; and

    (ii) Indemnification from the Government against environmental liabilities.

(f) An appropriate right-of-way to be acquired by the Government for the construction and operation of the onshore pipeline.

The Site. Given the magnitude of the project and the prior usage of the Site, it was considered reasonable that the financiers would expect some degree of indemnification against claims arising out of the occurrence of hazardous materials. Thus, as a requirement for project financing, Atlantic 1 was granted an indefeasible title to the Site[13] and the provision of environmental indemnities against claims arising from pre-existing conditions or activities. The scope of these indemnities covered the land site as well as the sea bed, harbour, foreshore, and other marine areas.

Upon the payment of U.S.$9,000,000.00, the Government agreed to take the necessary steps to compulsorily acquire the Site and grant, or procure the grant, to Atlantic 1, the formal documented legal rights to the Site (including the fee simple), Southern Corridor, Atlantic Harbour, and related areas. The Government also agreed to acquire the formal documented legal rights to sufficient lands not already owned by the Government, which would facilitate the construction, operation, and maintenance of an onshore pipeline "from, at or near Beachfield or Rustville" to Point Fortin.

Atlantic 1 Undertakings. Atlantic 1 gave undertakings that included the following:

(a) To procure that its general contractor for the LNG Facility did not cause large segments of the LNG Facility to be pre-assembled outside of Trinidad and Tobago save for traditional vendor packaging and normal prefabrication of steelworks and piping.

(b) To provide financial support for the founding of a Skills Development Centre, to be established by NGC on behalf of the Government, for the purpose of developing skilled

---

[13] "Site" is a defined term that meant certain lands comprised of the Reclaimed Lands, the Point Ligoure Road Frontage Area, and lands formerly occupied by the refinery.

labour consistent with the needs of energy and energy-related industries in Trinidad and Tobago.

(c) Following the investment decision and the establishment of the Centre (of which Atlantic 1 was to be a trustee), to pay the trustees the sum of U.S.$8,000,000 and a further sum of U.S.$5,000,000 to the trustees in annual installments over a period of 20 years.

(d) To institute a program of assignment and training within Atlantic 1 so as to provide training for at least 20 persons in various disciplines (but not more than 10 persons at any one time) designed to promote technology transfer through the phases of engineering design, construction, commissioning and operation of the LNG Facility.

(e) To provide in the EPC Contract that preference would be given to qualified companies, firms, and persons, residents in Trinidad and Tobago, who met the quality, cost, and schedule requirements of the Project in the award of sub-contracts with significant emphasis and investment in the training of nationals of Trinidad and Tobago.

(f) To achieve a minimum of U.S.$100,000,000 of local content within the EPC Contract. In that regard, a mutually agreed measurement system, monitored jointly by the Government and Atlantic 1, was to be developed and periodic reports issued to track progress. In the event that this commitment was not met, Atlantic 1 was obligated to make an additional contribution of U.S.$3,000,000 to the Centre, paid to the trustees of the Centre at the end of the first financial year of the Company after the Production Day.

(g) To pursue opportunities to increase the local content to a preferred level of U.S.$150,000,000 within all of Atlantic 1's contracts for the design, procurement, construction, commissioning, and start-up of the LNG Facility and the associated services in relation thereto but not so as to increase the overall cost of the LNG Facility.

(h) To continue good faith negotiations with PPGPL to seek a mutually acceptable agreement regarding the sale by Atlantic 1 to PPGPL of the NGLs manufactured by the LNG Plant.

(i) To design the LNG Plant to include the technical capability for the installation of the necessary equipment to extract ethane and to consider, in good faith, any Government initiatives for the commercial extraction of ethane, taking into consideration Atlantic 1's contractual obligations at the time.

Ethane Removal. The amount of ethane to be made available from the Train 1 Plant had been constrained by the specifications in the LNG sales contract between Atlantic 1 and Enagas. This contract specified that Atlantic 1 would not sell LNG to Enagas with a gross heating value of less than 1030 BTU per scf and a density of less than 431 kg/m^3. The high density and calorific value specifications in the Enagas contract had arisen because the terminals at Barcelona and Cartagena, which would receive LNG from Trinidad at the time, had received LNG supplies with high BTU content from Libya and Algeria. If high- and low-density streams of LNG were mixed, in certain circumstances, it would cause the LNG tanks to "roll over." The problem of "roll over" could have been removed by the installation of mixing facilities at the terminals and there were encouraging signals from Enagas that they would later relax the BTU and density specifications.

LNG Expansions. Under section 3.11 (d) (iv) (A) of the Shareholders' Agreement, any "Expansion" was regarded as occurring within a division of Atlantic 1. Since, at the outset, the economics and other relevant circumstances for exercising ministerial discretion to grant incentives for LNG Expansions would not have been known, the Order granting incentives in respect of Train 1 expressly excluded LNG Expansions.

Ownership of the Onshore Pipeline. In the Atlantic LNG forum, NGC had formally objected to Amoco's proposal to build and own the pipeline from Abyssinia to the LNG plant at Point Fortin. The Amoco proposal had been approved by the Board of Atlantic 1, ostensibly, due to the imminent deadline, 30th September 1995, for executing the gas supply contract (a condition precedent in both the Cabot and Enagas LNG Offtake contracts). NGC had been unwilling to allow Amoco to own and control the onshore pipeline from Abyssinia to the LNG plant at Point Fortin because, in principle, Amoco could have later constructed a pipeline from Picton to Point Lisas to supply gas to NGC's consumers at Point Lisas.

If Amoco got a foothold as an owner of a land-based pipeline, with such ownership, World Bank and other pressures might have made it more difficult for the Government to retain NGC's monopoly as a gas transporter. On the other hand, the Government's consent for Atlantic 1, a diversely held company, to own the onshore pipeline for the specific purpose of serving its own needs, i.e., the production of LNG, was perceived by NGC to be far less complicated. On that basis, NGC had agreed to Atlantic 1's ownership of the onshore pipeline under the terms of the Shareholders' Agreement.

The Pipeline Agreement. During the Hilton negotiations, NGC and Amoco agreed the following very broad principles, which were subsequently implemented in a formal agreement:

(a) *Amoco will retain full liability for its gas supply obligation to Atlantic 1.*

(b) *Amoco will conclude negotiations with NGC for the control, design, procurement, construction, and operation of a pipeline system comprising –*

    i.   *A 40-inch offshore pipeline from Amoco's offshore East Mayaro field to Abyssinia on the South-East coast of Trinidad.*

    ii.   *A 36-inch pipeline from Abyssinia to Point Fortin, dedicated to supply feed gas to Atlantic 1 (the offshore and onshore pipelines being collectively referred to as the "Pipeline System").*

    iii.   *Both parties contemplated that the offshore portion of the Pipeline System would be constructed by July 1998.*

(c) *Amoco will construct the Pipeline System, but NGC will have the right to receive and review all design data.*

(d) *Amoco will, at its own cost, operate and maintain the Pipeline System and will be liable for any loss or damages arising from or in connection with the Pipeline System.*

(e) *NGC will make a capital contribution (U.S.$ 53 million had been proposed, in installments) towards the Pipeline System, that contribution to be allocated to NGC's share of the costs of the onshore portion of the Pipeline system.*

(f) *In consideration of NGC's capital contribution payment to the Pipeline System, NGC will own the onshore portion of the Pipeline System.*

(g) *NGC will have the right to obtain tax benefits in respect of its capital contribution allocated to the onshore portion of the Pipeline System, and Amoco will have the right to obtain tax benefits in relation to its expenditure on the Pipeline System.*

(h) *NGC will lease the onshore portion of the Pipeline to Amoco for U.S.$ 1.00 per year and will pay a tariff to Amoco of U.S.$ 1.00 per year for the capacity rights for a period of 99* *years in the offshore portion of the Pipeline System. The respective periods of the lease and licence were to be negotiated.*

(i) NGC will retain any condensate associated with any gas owned and delivered by NGC into the Pipeline System.

(j) NGC will compensate Amoco for any losses resulting from the co-mingling in the offshore portion of the Pipeline of the higher calorific gas dedicated for LNG with that required by NGC for domestic consumers.

The agreed Pipeline Principles were beneficial to NGC because -

(a) NGC remained the sole owner of onshore pipelines and was able to ensure that any gas delivered through the LNG pipeline would be used exclusively for manufacturing LNG.

(b) There were concerns about possible damage to the onshore portion of the Pipeline System close to the sea in the vicinity of the LNG Plant site at Point Fortin, and Amoco had retained liability for any loss or damages arising from or in connection with the Pipeline System.

(c) In order to meet additional demand at Point Lisas, NGC had to incur costs to expand its transportation system by building a new onshore pipeline from Abyssinia to Point Lisas and an offshore pipeline. These pipeline costs were estimated to be U.S.$ 115 million, with the onshore section costing U.S.$ 36 million. As a result of the Pipeline Principles, NGC's pipeline capital costs to meet gas demand had been reduced to U.S.$ 53 million.

The Gas Supply Principles. The key areas of the agreed Gas Supply Principles were:

(a) The Gas Sales Contract dated 25th November 1991 ("the 1991 Contract") between the parties would be amended with effect from 1st January 1996 and would incorporate the Gas Supply Principles in all material respects.

(b) A retroactive adjustment of the prices paid for gas delivered during 1996 would be made after the 1991 Contract had been amended.

(c) The Initial Term of the 1991 Contract would be extended from 1st January 2011 to 1st January 2019.

(d) 1996 Contract Extensions. The Parties would begin discussions regarding the Trinidad gas market (excluding Atlantic LNG) and Amoco's gas reserves during the 1st quarter of

the Contract Year beginning 1st January 2014. The Contract would be extended beyond 2019 in accordance with specific requirements.

(e) Quantities. NGC could request increases in the Daily Contract Quantities (DCQ) and corresponding DCQ ceilings up to the following DCQ and DCQ ceilings:

Available Period for Max DCQ	Max DCQ and DCQ Ceiling
Present – 1st July 1998	350 MMscfd
1st July 1998 – 1st January 1999	500 MMscfd
1st January 1999 – 1st January 2019	700 MMscfd

(f) Gas Pricing.[14] Provisions were made for different pricing terms in relation to various Tranches of gas, and the first 350 MMscfd was characterised as the 1st Tranche.

(g) Reserves. Amoco was required to provide NGC with confidential, independent reserve audit reports, but such reports were not to be interpreted or to be construed as a guarantee or warranty of the audited gas reserves or to be interpreted or construed to amend Amoco's contractual obligations.

(h) Conditions Precedent. The Gas Supply Principles and the final amendment to the 1991 Contract were subject to –

    i.   the Government granting Amoco additional 10-year extensions of all of its Exploration and Production licences beyond the 15-year renewal period; and

    ii.   The execution and effectiveness of agreements between NGC and Amoco that were contemplated by the Pipeline Principles.

Project Authorisation. Under the terms of the Shareholders' Agreement, various conditions had to be satisfied prior to Project Authorisation. One such condition was for the Government and Atlantic 1 to enter into a mutually acceptable Project Agreement for Train 1. When the resolution of the issues raised during Hilton negotiations was imminent, the parties arranged to have a signing ceremony with the Prime Minister at the Point Fortin site. This ceremony was attended by all the directors of Atlantic 1. Prior to the signing ceremony, a meeting of the Atlantic 1 Board was

---

[14] The specific commercial pricing terms have been excluded.

convened in a minibus on the site and adjourned to permit the formal signing of various agreements. After the ceremony, the Board meeting was reconvened in the minibus, and the Project was formally approved by the Board in the presence of Officers of Atlantic 1 and shareholder representatives.

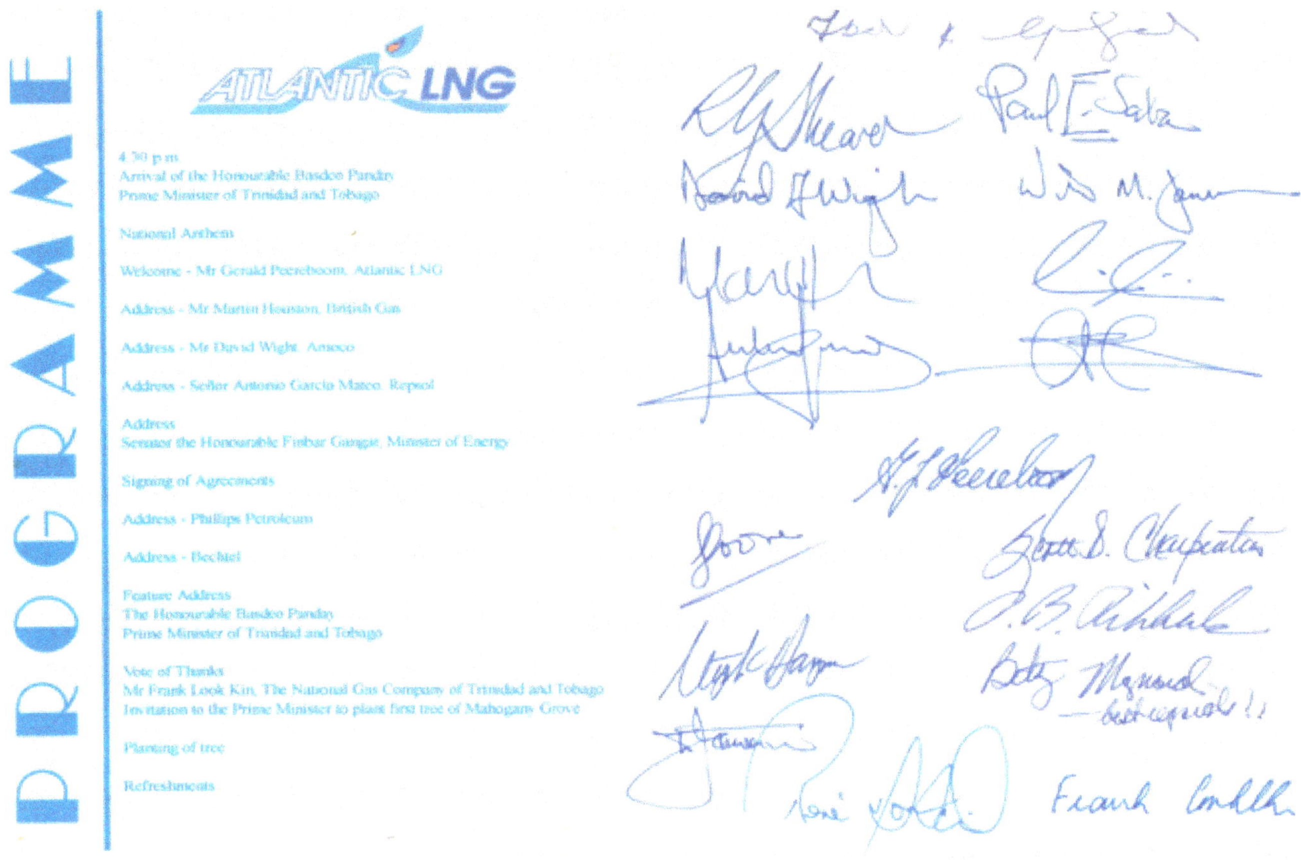

Sponsor Approvals. Prior to the award of the EPC contract, Atlantic 1 was also required to obtain satisfactory financial commitments from its Shareholders in order to consider and, if appropriate, authorise, the relevant budget for awarding the EPC contract. Thus, under the terms of the Shareholders' Agreement, each Shareholder was required to determine whether it wished to continue its participation in the Project before Atlantic 1 approved the budget for the award of the EPC contract. At the time, NGC had committed approximately U.S.$ 4 million as part of its share towards Project costs, and it was necessary to obtain NGC's Board approval for continued participation. In my report to the Board seeking its approval (which was granted), the Board was informed (among other things) that –

(a) The Stone and Webster due diligence Report had confirmed the Project to be very robust.

(b) An independent audit of the gas reserves dedicated to the Project had confirmed that the reserves were sufficient to enable Amoco to supply the volumes contracted to Atlantic 1.

(c) Project completion was facilitated by –

    i.   A Sponsor completion guarantee (on a several basis) during the construction period and

    ii.   Cabot LNG's commitment to make further investments in its existing LNG assets.

(d) The anticipated return on equity to NGC was *circa* 18 %.

Implementation of the Gas Supply Principles. As the Vice President, Business Development, I subsequently had the responsibility to implement the Gas Supply Principles into a formal contract. Under the terms of the Gas Supply Principles, Amoco and NGC were required to negotiate, in good faith, to finalise and execute a detailed amendment to the 1991 contract, to be effective January 1, 1996 (the "1996 contract"), incorporating in all material respects the contents of the Gas Supply Principles.

The Negotiating Background. The consolidation of the Gas Supply Principles and the 1991 Contract had to be accomplished against the following commercial background:

(a) *The Arrangements with Atlantic 1*. In the feedstock purchase agreement that was executed between Atlantic 1 and Amoco, Amoco had given Atlantic 1 many legally binding feedstock gas supply security commitments, including:

    i.   The dedicated gas fields would contain sufficient reserves to enable Amoco to perform its obligations under the contract;

    ii.   Amoco would evaluate probable and possible reserves to provide upside to the then-existing levels of proved reserves dedicated to the contract; and

    iii.   Any loss or failure of the relevant gas reservoirs and the deliverability associated therewith due to natural depletion, or the absence of economically recoverable gas (other than such an absence which itself is the result of *Force Majeure*), was not considered to be an event of *Force Majeure*.

(b) *The 1991 Contract with NGC*. In the 1991 Contract between NGC and Amoco –

    i.   Amoco had dedicated to the Contract all associated gas reserves and all economically recoverable gas well gas reserves that could have been produced from wells drilled from Amoco's existing platforms. At the time, the reserves that could have been produced from Amoco's existing platforms would have satisfied NGC's requirements for 350 MMscfd under the terms of the 1991 contract.

    ii.   Amoco had the discretion to dedicate gas from other parts of its licensed area. This discretion was commercially acceptable to NGC because, as the only buyer of gas from Amoco, Amoco had an incentive to maximise sales to NGC for the only natural gas market, namely, NGC's domestic market.

However, with the advent of Atlantic LNG in 1995, the commercial landscape for natural gas supplies had changed when the Gas Supply Principles were being implemented in 1996:

(a) There were now two buyers of gas from Amoco. Thus, it could not be precluded that Amoco would now exercise its discretion differently (with a strict interpretation of its legal commitments to NGC when undertaking its obligations under the Gas Supply Principles).

(b) The 1996 Contract had contemplated a long-term gas supply arrangement until 2019, at the earliest, which meant that NGC had to ensure, at the outset, that the proved reserves dedicated to the 1996 Contract had provided sufficient gas supply security for NGC's current and projected projects.

The difficulty was that the Gas Supply Principles had left it open to interpretation as to which terms in the 1991 Contract should **not** be amended when implementing the 1996 Contract.

Amoco's view was that all the terms of the 1991 Contract should apply, save, and except for the changes that were "required by" the Gas Supply Principles, and where those Principles were silent, the existing terms of the 1991 contract should apply. However, Amoco's interpretation begged the question as to what changes were "required by" the Gas Supply Principles.

NGC, with a different perspective, had contended that the terms of the 1991 contract should apply unless the context had required otherwise and, with the advent of Atlantic 1, in NGC's view, the context had required "otherwise." These differences in perspectives led to protracted negotiations although, during the negotiations, NGC had slowly been able to obtain the following beneficial concessions from Amoco (which were outside the scope of the Gas Supply Principles):

(a) An upward adjustment in the maximum DCQ to bring the DCQ to 700 MMscfd.

(b) The flexibility to use up to 450 MMscfd in the **offshore** system to be constructed by Amoco, pursuant to the Pipeline Principles, even though the Gas Supply Principles had specifically indicated that the First Tranche of gas (350 MMscfd) should utilise the existing **onshore** delivery point in the 1991 Contract.[15]

(c) The ability to reduce the DCQ from 700 MMscfd to 450 MMscfd in the short term without a take-or-pay penalty.

Ultimately, the major roadblocks related to (a) Reserves dedication and (b) Reserves Risk.

Dedication of Reserves. The gas reserves referenced under the 1991 Contract would have been approximately 2.7 Tcf (according to an audit report prepared by Gaffney Cline). However, approximately 5.4 Tcf of gas had now been required in respect of the volumes to be delivered under the 1996 Contract. Despite the doubling of the DCQ (from 350 MMscfd to 700 MMscfd), Amoco had argued that the 1996 Contract did not impose a legal obligation upon them, the gas supplier, to dedicate more reserves than those previously dedicated under the 1991 Contract.

NGC contended that this approach was, at face value, fundamentally flawed: how could Amoco double the volume commitment for the gas supply to NGC without increasing the quantity of dedicated reserves in respect of that supply commitment?

Reserves Risk. Amoco, as a "reasonable prudent operator," was only prepared to prove up more reserves (and incur additional expenditure for the development) after nominations for gas by NGC.

---

[15] NGC's technical evaluation was that the use of the new offshore pipeline, as proposed, was critical to enable NGC to meet the projected demand during the middle of 1998 and 1999 whilst permitting a sufficiently high delivery pressure at PPGPL.

That position was not unreasonable because 'just-in-time' capital expenditure would have assisted any gas producer in obtaining a better return in respect of the development of gas reserves. Thus, in that regard, NGC was willing to compromise.

However, the contentious issue was that even after any such NGC nomination, Amoco was unwilling to guarantee that the development would have provided sufficient gas because Amoco was not prepared to warrant the **existence** of the reserves. In essence, the gas producer, the Seller (whose core business was to explore, develop, produce, and sell gas), was asking NGC, a Buyer, to take the risk – the Reservoir Risk - that the Seller's gas reservoirs would be able to produce the gas. In January 1997, I formally expressed my concerns to my Amoco counterpart in Houston in the following terms:

> *"Dear Sir,*
>
> ### *Gas Supply Negotiations*
>
> *We refer to our discussions on the captioned subject matter and, in particular, to our letter of 12th November 1996, which offered a proposal for the resolution of the outstanding issues. As matters stand, the biggest issue relates to Amoco's obligation to supply gas to NGC.*
>
> *Amoco has expressed a significant concern that NGC is attempting to renegotiate the 1991 Contract to obtain a 20-year warranty of gas supply. The point relates to the concept of a "reasonable prudent operator," which applies in the 1991 Contract and also to reserve risk. The supply obligation is important to NGC because NGC has to consider a 20-year contract.*
>
> *NGC would be expected to agree that capital spending may take place on a phased basis in accordance with an agreed development programme. .... In that regard, NGC is prepared to compromise and accept a supply obligation that takes into consideration the interests of both Amoco and NGC: a reasonable expectation of profit for Amoco (on measurable criteria) and sufficient express safeguards that would not prejudice NGC's ability to promote, develop and sustain gas-based industries.*

*For example, we would have to be satisfied that Amoco is **not** permitted to decline to supply gas to NGC on the basis that Amoco can obtain a higher price from a third party or a better investment elsewhere. The standard of supply by Amoco to NGC must also be one of good faith and must not be reduced by obligations that Amoco may have to other parties.*

As to reserve risk, we believe that this is an area where, in principle, both parties' interests should be aligned because, as you know, the existence of reserves is a growing concern for financiers of new investments. In this context, Amoco has only offered to implement a programme underlined designed to increase the quantities of gas in the context of nominations for gas by NGC. Accordingly, Amoco's approach presents a weakness in marketing gas to potential new customers because an obligation to put in place a programme is clearly not an obligation to supply the increased quantities of gas. Notably, Amoco has given Atlantic LNG a firm commitment to sell the specified quantities of gas. ………

We have now offered an even further compromise solution to Amoco's obligation to supply gas above 350 MMscfd as part of the "package." However, we are unable to accede to your request for changing to a new method of measurement, although we are prepared to work with you, in good faith, to see how we could accommodate your desire for greater operating flexibility. ………

Yours faithfully,

Dr. René L Monteil
Vice President, Business Development"

Amoco had requested NGC to change the method of measurement for gas in the 1991 Contract in order to dovetail with the metering arrangements that had been agreed earlier between Amoco and Atlantic 1. Since it would have been more practical for Amoco to have one system of metering the gas that was delivered to both Atlantic 1 and NGC, NGC had been holding out on that issue as leverage to obtain concessions from Amoco.

The Gas Supply Principles provided improved gas pricing terms for NGC, but these improved terms were only triggered after the 1996 Contract was executed. Thus, on 20th June 1997, almost exactly 12 months after the Gas Supply Principles were signed, I provided NGC's Board with an update on the negotiations and sought (and obtained) the Board's approval for the management to execute the 1996 Contract. It was also recommended that NGC defer the argument on the adequacy of reserves to "another forum" when LNG Train 2 approval was under consideration.

NGC ultimately executed the 1996 Contract.

IN WITNESS WHEREOF, this Gas Sales Contract is executed by the duly authorized representatives of the Parties hereto.

Signed by **Frank Look Kin,**
President, for and on behalf of
THE NATIONAL GAS COMPANY OF
TRINIDAD AND TOBAGO LIMITED
in the presence of:

Signed by **David G. Wight,**
President and General Manager,
for and on behalf of **AMOCO TRINIDAD**
OIL COMPANY
the presence of:

Chapter 5: LNG Expansions

Train 2. Train 2 raised various strategic issues for NGC, which are reflected in my 1997 strategy note as Vice President, Business Development:

"The decision of the Board of Atlantic LNG Company of Trinidad and Tobago ("Atlantic") to set up a team to investigate the options for building an additional liquefaction Train (an "Expansion Train") raises interesting strategic issues for NGC.

Gas Reserves. NGC has a duty to ensure that reserves are not exported to the detriment of its customers and that the reserve base is sufficiently strong to support future petrochemical growth. Our base case demand projections (which take into account existing and likely contracted supplies) show that the growth in demand could reach 1,200 MMscfd by the year 2005 if we do not lose any existing customers. This demand would require about 9.3 Tcf for the period until 2015, even if we were to assume no further growth beyond 2005. However, 4.8 Tcf of the 9.3 Tcf total proved reserves (excluding the North Coast Marine Area) are already dedicated to first-train LNG.

Accordingly, as a first position, I think that NGC should have priority over available proved reserves and be in no worse position than Atlantic with respect to domestic reserves for an Expansion Train. Even so, only 4.5 Tcf of proved reserves (a shortfall of 4.5 Tcf) will be available for NGC's customers for the period until 2015, until this reserve picture changes with additional development and further discoveries.

But as matters stand, the reserve base may not be attractive to investors and financiers in a project such as an aluminium smelter which will require reliable gas supplies over an extended project life of at least 20 years. In principle, therefore, the presumption should be against using existing domestic reserves for an Expansion Train.

The Atlantic Forum. *It is useful to appreciate some of the relevant provisions that govern how Atlantic would pursue Expansion Trains even though discussions are at an embryonic stage and Government approvals have not formally been sought.*

By section 3.11 of the Shareholders' Agreement, voting shareholders holding at least 51% of the relevant shares (Amoco and British Gas would reflect 60%) may propose that Atlantic build an Expansion Train by submitting a budget for the Expansion. Thus, NGC, even with support from Cabot (and Repsol if it has not already aligned with Amoco), would be unable to block such a proposal at a shareholders' meeting. Moreover, Amoco and British Gas, under the terms of a letter agreement with Atlantic, each have preferential rights to supply to Atlantic up to 50% of the gas for an Expansion Train (and any subsequent Train). They may either submit a joint proposal for a 100% supply of gas for the Expansion and conduct negotiations with Atlantic jointly; alternatively, each could negotiate separately with Atlantic, on whatever terms, with respect to its 50% portion of the gas for the Expansion.

In theory, the directors of Atlantic stand in a fiduciary relationship with the company as principal, and this fiduciary relationship imposes upon them duties of loyalty and good faith to the company. Whilst a "company" is usually defined in Equity by reference to the Shareholders as a whole (and not by reference to the "company" as an entity distinct from its members), this does not preclude the need for the directors to consider the interests of the "company." Arguably, the Directors should first be satisfied that Bechtel will construct the first LNG Train within the scope of the agreed terms of the EPC contract before committing to another contract with Bechtel for an Expansion Train.

Also, before a proposal is made for an Expansion Train, Atlantic's Board must have approved an EPC contract for the expansion, and Atlantic must also have entered into LNG Purchase and Sales Agreements and Feedstock Purchase Agreements for the Expansion. It is not clear how long these agreements would take to negotiate.

The reality is that, in practice, the vested interests of the majority of Atlantic's shareholders have shown to be paramount for Atlantic's directors in making decisions. This stems from the provision in the Shareholders' Agreement that permits each Shareholder to grant or withhold its consent, approval, or vote, in its sole discretion, "without regard to the interests of or any obligation to the other Shareholders or the Company, it being understood that no shareholder has any fiduciary or other expressed or implied duty to act in the best interest of the other Shareholders or the Company" (Section 14.1). It would require extraordinary mental gymnastics for the directors to make a distinction between a Board meeting and a Shareholders' meeting.

The Way Forward. *However, there might be overriding strategic benefits to allowing the momentum to be maintained for Expansion Trains because, in so doing, it could facilitate the introduction of alternative gas supplies from the Plataforma Deltana region in Venezuela. LNG might, possibly, be a carrot for the Venezuelans, which could lead to arrangements with NGC for additional supplies to maintain the development thrust. Admittedly, there are complex and possibly time-consuming issues to be resolved with respect to extraterritorial gas from Venezuela (which will be discussed elsewhere).*

In any event, in any negotiations on behalf of Atlantic with other gas suppliers, NGC will have a significant role to play because Atlantic cannot be represented by Amoco and/or British Gas, who each will have a declared interest in Feedstock Purchase Agreements for the Expansion.

Clearly, the skill will be how to maneuver to achieve NGC's objectives."

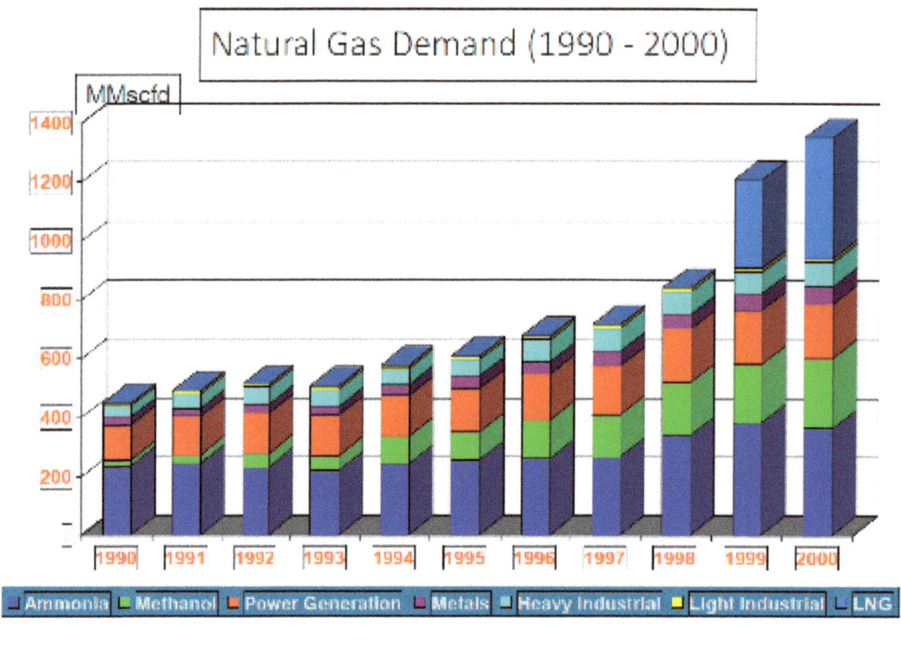

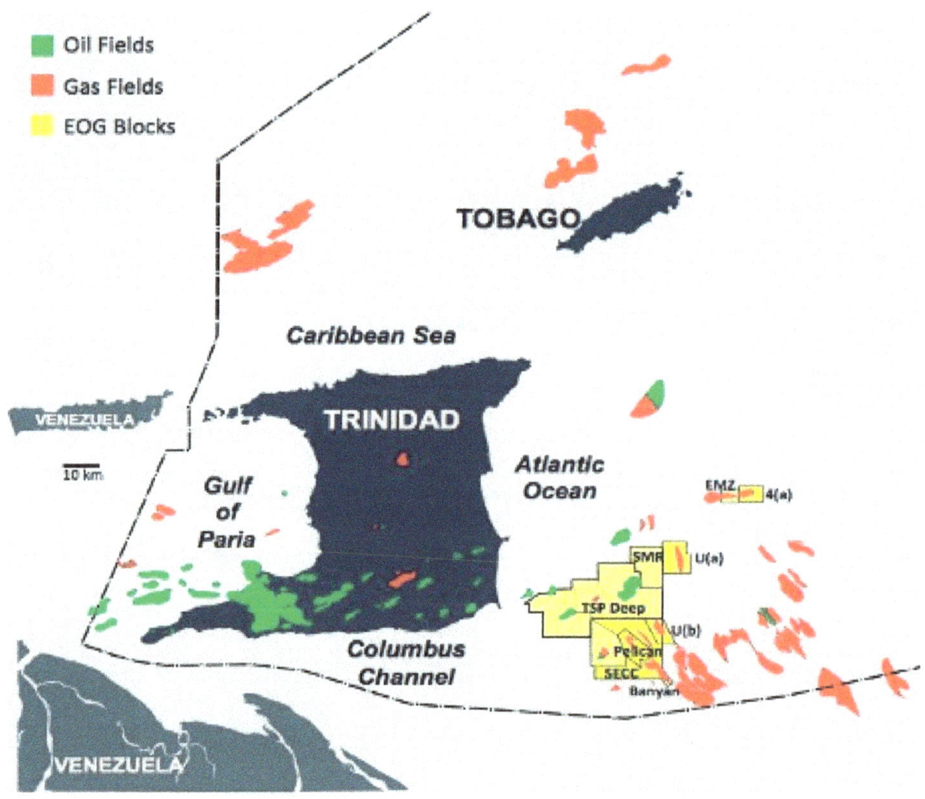

At the time, the gas reserves in the NCMA, without infrastructure, were stranded, and it was necessary to produce large volumes of gas from this area so that these reserves and the associated pipeline infrastructure could be developed economically.

As to the introduction of alternative gas supplies from the Plataforma Deltana region in Venezuela, NGC had held discussions in Venezuela with PDVSA[16] pursuant to an initiative with Conoco to develop natural gas and related industries in Trinidad and Tobago.

Who should carry out the Expansion? Under the terms of the Shareholders' Agreement, as preconditions to a proposal for an Expansion Budget, Atlantic 1 must have first executed feedstock gas supply contracts and LNG offtake contracts for the Expansion. However, the Shareholders' Agreement was silent as to the mechanism for negotiating the LNG offtake and feedstock supply agreements for an Expansion. Notably, the Train 1 LNG offtake contracts with Cabot (Article 5.9) and Enagas (Article 5.8) had given these buyers preferential rights to negotiate first with Atlantic 1 for 60% and 40%, respectively, of the LNG produced from the Expansion. Thus, on one analysis, these negotiations should be conducted by Atlantic 1, particularly since Atlantic 1 ultimately had to execute the contracts. In the same vein, Atlantic 1 should negotiate the feedstock gas supply agreements for the Expansion because it had signed the Letter Agreement with BG and Amoco, which had given these gas suppliers preferential rights with respect to "Expansion Gas."

Another approach, advanced by Amoco, was that the LNG offtake and feedstock supply negotiations for the Expansion should be shareholder driven. That view appeared to have been based upon a requirement that the approval of the shareholders was required for the negotiation and execution of the contracts with a shareholder's affiliate (pursuant to Article 5.7 of the Shareholders' Agreement).

However, NGC believed that Article 5.7 of the Shareholders' Agreement neither precluded the negotiations being led by Atlantic 1 nor did it prescribe that the negotiations must be carried out by the shareholders. Rather, the approval of the Shareholders was merely a precondition to the commencement of the negotiation or execution of the Expansion Agreements.

---

[16] In a very cordial meeting with Juan Szabo, then the PDVSA Vice President Exploration and Production, at his office in Caracas, he told Frank LookKin and me that he had fond memories of Trinidad and showed us a picture on his wall of an exploration rig in Trinidad and Tobago's waters.

In any event, the issue as to whether the development of an Expansion Train should be carried out within ALNG or led by shareholders had raised important points of principle for NGC (as noted in my internal views on 11th August 1997):

> *"At the operating level, if development activity were to be carried out by shareholders, there would potentially be less opportunity for Trinidadians to gain experience with respect to such activity. Moreover, all employees and directors would have fiduciary duties to ALNG, which would not necessarily exist in a shareholders' forum. .... It is also expressly understood and agreed by section 14.3 of the Shareholders' Agreement that each party may continue and commence independent businesses, including those which compete with (or are similar to) the project in Trinidad."*

Tensions escalated among the shareholders during 1997 as the differences between their respective commercial objectives began to crystallise. The low point was a 3-day futile meeting among the Shareholders at Repsol's offices in Madrid.

Cabot. At the meeting in Madrid, Cabot suggested that Expansion Train discussions should take place in the context of a more clearly defined commercial structure for the Expansion. Cabot's rationale was that there were then *"obvious overlaps of shareholder interests giving rise to tensions that did not exist with Train 1."* In Cabot's view, as indicated below, there were:

> *"Many outstanding questions to be settled before moving ahead on an Expansion Train:*
>
> *(a)   Who would be negotiating?*
>
> *(b)   Could there be simultaneous negotiations of the Gas Supply Agreement?*
>
> *(c)   What is the role of the shareholders?"*

Thus, Cabot believed that triggering Article 5.9 of its LNG Sales Agreement, as proposed by Amoco, "would be premature."

Amoco. At the meeting, when pressed by Cabot as to whether Amoco saw itself as a buyer of LNG, Amoco's response was somewhat guarded but instructive:

> *"There is no clear-defined market that belongs to Amoco, but we are working in different places to obtain markets. If opportunities develop, we will inform ALNG."*

Thus, Amoco was already furthering its interests in forward vertical integration along the LNG value chain by becoming an offtaker of LNG.

Repsol. Repsol, who (in conjunction with Amoco) had the expectation of being a gas supplier to an Expansion Train, had also indicated that they (Repsol/Enagas) could take the full LNG market for the Expansion. Essentially, Repsol had now backwardly integrated along the Atlantic LNG value chain by becoming a potential gas supplier in Trinidad.

British Gas. The Letter Agreement with Atlantic 1 (which entitled BG to provide 50% of the gas supply to all LNG expansion Trains) did not serve Amoco's interests if British Gas had been unable to provide gas for an Expansion Train in a timeframe that would meet Amoco's objectives. Amoco, with sufficient proven reserves and available infrastructure, had a clear strategic advantage over British Gas and could move quickly to supply Expansion Gas to Atlantic 1. British Gas, with aspirations as a seller of Expansion Gas from the NCMA, which was still under development, had expressed some skepticism about the need to move quickly.

NGC. NGC was generally a moderating influence with the other shareholders. However, since a strategic objective for the Expansion was to develop the natural gas reserves from the NCMA, NGC tended to be aligned with British Gas on the timing of an Expansion.

Eventually, both BG and Amoco (later, BPAmoco) became fully vertically integrated by developing their interests in LNG receiving terminals.[17]

---

[17] In May 2001, the BG Group announced that it had signed an agreement to take all of the available capacity at the LNG importation terminal in Lake Charles, Louisiana for 22 years beginning in January 2002. From 1 September 2005, BG Group had the right to utilise 100% of the terminal's capacity.

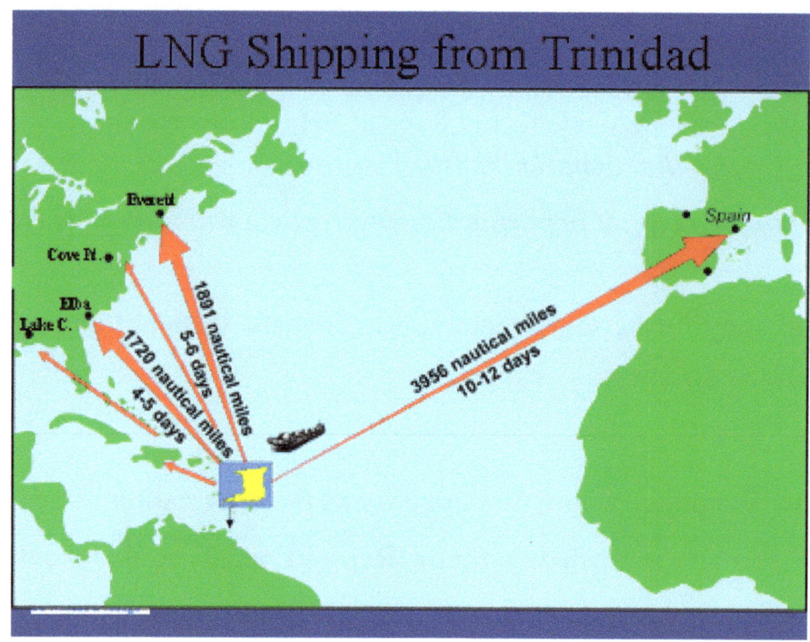

The Investor Expansion Committee. In January 1998, the shareholders entered into a cost-sharing agreement (the "CSA"), which established the activities and associated costs that were required before an Expansion Budget could be submitted to the Board of Atlantic 1 for approval. Under the terms of the CSA, an Investor Expansion Committee ("IEC"), composed of representatives of each shareholder, was constituted to provide for the overall direction of the work for the Expansion ("Expansion Work"). To further provide for the implementation of the directions of the IEC, a Project Development Team was established, which was also composed of representatives of each shareholder. The Project Development Team was required to work under the chairmanship of a Project Manager, who was appointed by the IEC. The Project Manager was required to initiate, supervise, direct and administer all the Expansion Work approved by the IEC or contemplated by work programs and budgets approved by the IEC.

When my contract with NGC came to an end at the end of February 1998, I took up the position, within Atlantic 1, as the Project Manager for Train 2. Although there was generally excellent cooperation among the members of the Project Development Team, the squabbling persisted within the IEC. Nothing of any real significance could be accomplished.

Eventually, Amoco's fast-track approach led to a warning letter from Cabot in July 1998:

"July 22, 1998

To the Officers, Directors, and
Shareholders of Atlantic LNG
Company of Trinidad and Tobago (the
"Company")and their respective
Affiliates

Re: Expansion of the Company

Dear Sirs:

By a separate letter of July 21, we have stated that the expansion proposal by Amoco and Repsol violates contractual and other legal obligations owed to the Company, its shareholders, and other interested parties. As indicated in that letter, we are willing to consider expansion in accordance with the existing contracts and agreements and applicable law. It is our right, as well as the right of each other shareholder, to require that the expansion be so conducted.

However, while reserving all of our and the Company's rights under the existingcontracts and agreements and applicable law, we are also willing to consider a compromise expansion proposal in an effort to achieve a reasonable accommodation of theinterests of all parties. Absent such compromise, we believe the present dispute regarding expansion is likely to result in major litigation involving Cabot, Atlantic, and one or more of Atlantic's shareholders. In the interest of averting such litigation and resolving that dispute, we are prepared to discuss the attached outline proposal.

If there is interest in pursuing such discussion or any other compromise or settlement discussion, we are willing to do so on the understanding that any such discussion does not constitute a waiver of, or compromise, any rights or obligations of any party under existing agreements and applicable law, and that any compromise expansion proposal, to the extent it departs from existing contractual provisions, can be effected only with the agreement of all interested parties.

Very truly yours,

141

CABOT TRINIDAD LNG LIMITED	CABOT LNG CORPORATION
By: _RlfShearer_	By: _RlfShearer_

July 22, 1998

For Settlement Purposes

Attachment to Cabot Letter of 22nd July, 1998

Proposal for Atlantic LNG
Expansion

Single Expansion Train: *The parties would initially pursue a single expansion train. The schedule would target a year-end investment decision and a late 2001 commercial service date. Any additional expansion would be considered following a final investment decision on Train 2.*

Tolling Facility: *Train 2 would be built and operated by Atlantic as a processing/tolling facility. BG and Amoco would each contract with Atlantic on a long-term "hell or high water" basis for the right to have gas processed and liquefied at the facility. BG and Amoco would each commit to half the capacity of the facility (on MMBTU basis).*

Tolling Fee Design: *The tolling fee would have 3 parts: (i) a fixed annual fee covering Atlantic lease payment (see below), (ii) a fee designed to cover Atlantic's incremental operating costs resulting from the Train 2 facilities, paid·as a function of availability, and (iii) a variable charge paid per MMBTU delivered (primarily fuel retainage plus a $.05/MMBTU ($1998) margin).*

LNG/NGL Sales: *BG and Amoco would each individually contract to sell their respective portions of the LNG and NGL production on an FOB or CIF basis and on such other terms as they individually deem appropriate. Atlantic could install NGL processing facilities and charge a separate NGL processing fee to producer(s) to recover the costs.*

***Plant Financing:** The Train 2 facilities would be owned by a bank or other financing entity, and leased to Atlantic. The lease would be secured by the "hell or high water" BG and Amoco processing contracts.*

***Joint Use and Operating Agreement:** Since the facilities would be leased to or owned by Atlantic no such Agreement would be required."*

During 1998, in my capacity as the Project Manager, with single-point responsibility to the members of the IEC, I continued to write the shareholders as the squabbling continued:

To All Shareholders on 5th August

"Gentlemen,

As you know, the effect of each side's rejection of the other's budget proposals for the expansion means that some meaningful work for the Expansion is not being pursued. It is clearly in the interests of everyone that we resolve differences quickly if the project is to have any chance of meeting a start-up date in January 2002. As matters stand, even that target looks very optimistic.

Notably, it was the ability to play on each other's strengths that enabled the shareholders to demonstrate a successful project and develop a reputation for Atlantic LNG. It seems not too long ago we were celebrating project approval in a mini-bus on the Point Fortin site!

The Expansion clearly raises many issues both of a legal and commercial nature, including, for example, the rights of shareholders under the terms of the current Shareholders' Agreement. Indeed, that Agreement already contemplates a mechanism for compensating a "Non-Investment Shareholder" with respect to a single train expansion within Atlantic LNG. It might be possible to build on this structure.

Since all parties have now indicated the willingness, in principle, to pursue at least a single train expansion, one way to break the current impasse would be to move ahead on the common ground. The parties could commit to a single train expansion and also commit to work, in the utmost good faith, towards an additional train. There is scope for many variations on this theme including time limits and compensation. The key is flexibility; the difficulty, of course, is a perception of the loss of commercial leverage by compromise.

But everyone loses if the gas stays in the ground.

I am therefore suggesting that, at the very least, we continue activity where there appears to be a common interest such as, for example, the work on the FEED phase for Train 2. Indeed, there will be many issues such as financing or a shareholders' agreement that are inherently time consuming.

In any event, it would be prudent for the parties to meet as soon as reasonably practicable to see how best to advance matters."

To Cabot on 6th August

"Gordon,

The difficulties encountered by Amoco would appear to be self-inflicted because they have taken steps without consulting the other shareholders. The problem, though, is that any BG/Cabot/NGC proposal would also be doomed to failure without support from Amoco. A key consideration, therefore, is what tactics should be adopted.

Amoco has the most to lose by a delay: there would be a significant loss in revenue by failing to monetise reserves and perhaps even undesirable consequences regarding the Repsol deal. However, a compromise, as I suggested in the note yesterday, would mean some delay which could, in principle, let in BG as a supplier for Train 3. The loss of opportunity to exploit gas reserves for half a train and its associated condensate (Mega bucks!!) will clearly be resisted by Amoco. Of course, taking into consideration the background and the agreements negotiated, BG can justifiably argue that it ought to be given some time to be in a position to supply Train 3. They, like Cabot, quite properly ought not to be "hustled out" of their rights.

I have some recollection of the Letter Agreement with ALNG that provides both BG and Amoco with preferential rights to negotiate gas supply with ALNG for Expansion Trains. I believe that it was a 50:50 deal for all Expansions. However, I do not recall anyone either discussing the timing between Expansions or a requirement that both parties must, in fact, be in a position to supply Expansion Gas in order to start the negotiation process with ALNG. Can one party make a pre-emptive strike?

On the one hand, it might be extreme for Amoco to move now to Train 3, in the context of knowing that BG is not even in the starting block for Train 2. But it would be equally extreme, in my view, for BG to have an open-ended restraint on Amoco until such time as BG considers itself in a position to supply Train 3 (or subsequent Expansion Trains). What happens if BG never finds gas for Train 3? The question is how long would be reasonable.

If the Amoco reserve base is as strong as David is saying (and he is not yet exposing his hand), I could see pressure being brought to bear that delay should not be permitted merely to give BG a potential opportunity to supply Train 3 (which may or may not be realised). It is also conceivable that Amoco would get political sympathy if he is able to placate NGC on reserves, the gas pricing to Spain is acceptable in the light of alternatives, and BG has the opportunity to supply half of Train 2. In any event, BG is not likely to kill the goose because of their interest in supplying Expansion Gas, and they could conceivably negotiate an option to supply Train 3 to be exercised by a prescribed time. I anticipate that, with time, there will be movement by the gas suppliers toward each other.

The market options on price and timing and crucial information (such as capital costs) are clearly important for a proper determination by all shareholders with respect to Train 3. Thus, from a practical point of view, I think the step-wise approach makes more sense. In any event, as I see it, in terms of a deal structure for Train 2 (and/or Train 3), Amoco is more likely to perceive BG as a greater threat than Cabot because of the capacity constraints at Everett and the uncertainty regarding LNG supplies to Brazil."

On Tuesday, 11th August 1998, BP announced a merger with Amoco, which further stalled the project.

Eventually, meetings of the shareholders were arranged to take place in Boston in September, and as the Train 2 project manager, I continued to write the shareholders.

To NGC on 2nd September 1998

"Frank,

I am writing you under a separate cover to give you my thoughts on the Expansion issues, some of which you may have already considered.

My major concern is with respect to the tolling arrangement, which at face value, does not appear to be of strategic benefit to T&T. There is a related issue on the question of the party who should carry out an expansion: should this be ALNG or should an expansion be shareholder driven? These are my concerns:

i)   The proposed arrangement will permit control of the project to be transferred from within ALNG to the suppliers, with the protection for other shareholders being limited to the negotiation of a processing fee. Thus, with Amoco, Repsol, and BG (80% of ALNG) having an interest in the supply, it is critical to determine how, in what circumstances, and by whom, any processing fee will be negotiated. Clearly, there are opportunities for conflicts that have to be resolved.

ii)   The Expansion could be pursued within ALNG with support from shareholders, as was the case with Train 1 development, particularly since a shareholder-driven Expansion has shown that there is less opportunity for harmony.

iii)   A supplier-driven Expansion ensures that the employees of the suppliers (as opposed to locals) get training in a critical aspect of the LNG business, namely, marketing. Trinidad should have a strong voice in this matter, and there is reduced opportunity for job creation and training of locals if the Expansion is no longer controlled by ALNG, i.e., by all the shareholders, including NGC. This should be nipped in the bud now, and the project should be developed, at the outset, in a manner consistent with National interests.

iv)   *The marketing of significant volumes of gas and the timing of revenue streams from gas monetisation will be controlled by suppliers (and not by NGC on behalf of GORTT). This has wide implications that can become a matter of significant concern because there is potential for the suppliers to use leverage on the Government if the Government revenues are in decline. Once the door has been opened and the principle accepted, it is difficult to predict the consequences, particularly in an ever-changing landscape.*

v)   *There is the possibility of pricing arrangements with external markets to favour suppliers, not the Government (who will derive most of its revenue from the wellhead price).*

vi)   *In the case of ALNG, the marketing negotiations were carried out by a team of shareholders that included NGC. However, with the tolling mechanism, it will not be a simple exercise (taking into account confidentiality) to ensure transparency and optimum pricing of LNG to benefit T&T. Indeed, as events have shown, NGC, to date, has been unable to obtain commercial details of the proposed arrangement between Amoco and Repsol.*

vii)   *It might be somewhat of an academic exercise that the Government ultimately has the power to approve projects (with any terms and conditions attached thereto) if the officials have no real check on the integrity of pricing proposals submitted.*

I have other points of detail, so please let me know whether you wish any further information regarding the above."

To NGC on 9th September 1998

"Frank,

I spoke to Simon Bonini briefly last night as he was on his way to the U.K. and he gave me some insight on BG's current thinking with respect to the planned meetings in Boston next week. He told me that he was trying to reach you but was unsuccessful.

From our discussion, I understood that Martin Houston had been having discussions with Repsol and Amoco who maintain their position regarding the 2-Train as opposed to a 1-Train Expansion. The repeated Amoco concern has been why the shareholders would wish to give up some of the Spanish market. Of course the Spanish market can still be preserved if ALNG is the contracting party both for sales of LNG and for the purchase of the feedstock.

I think that the real issue is that Amoco has positioned itself strategically to make gas sales directly to the international market and will, in so doing, obtain income from such sales that would otherwise be shared by the other shareholders of ALNG. This becomes quite relevant if the proposed gas prices in Spain are low. Perhaps the financing considerations of the project will be important.

In any event, the impression I got from Simon is that BG has now not precluded going along with the Amoco/Repsol approach but wished to contact you in advance so that you would not be taken by surprise on any discussions on Train 3. I do not know whether Cabot is also on board on Train 3 discussions although Simon did confirm to me that Martin had spoken to Gordon about the possibility of BG's stance at the proposed meetings.

In one sense I am not altogether surprised that BG is now prepared to entertain the 2-Train Expansion proposal bearing in mind that BG, NGC, and Cabot collectively do not have the requisite votes to enable a single-Train Expansion Resolution to be passed. Moreover, since BG is a gas supplier, one would anticipate that in the final analysis BG would have a greater deal of alignment with Amoco than with Cabot (or even NGC). In this regard it is not inconceivable that conceptually the germs of possible 'deals' are already in place.

However, BG (unlike, as it would appear, Amoco and Repsol) are more conscious of project development without regard to guidance from the State and it would be prudent to anticipate that you will be asked to express some policy views on various aspects of the project. NGC's position on gas reserves and the tolling arrangement are obvious candidates. As to the latter, it would be useful to know whether there are, and if so, what, key policy considerations exist in other countries that have approved a tolling mechanism for the export of LNG.

My best wishes to you next week."

To Cabot on 12th September 1999

"Gordon,

The resolution, in principle, of matters between Cabot and the sponsors of the Expansion is a step in the right direction. However, the daily use of approximately 1.0 Bcf of natural gas for LNG raises, in my view, many national interest issues for Trinidad and Tobago.

The first consideration is whether the existing reserve base is sufficiently strong to justify a 2-Train expansion. In this regard, I expect that NGC will wish to ensure that reserves will not be depleted to the detriment of its customers and that the reserve base will support optimistic projections for future petrochemical growth.

In a climate where money is tight, it would be prudent for NGC to be conservative with respect to available proven reserves even if there is cause for optimism with respect to "exploration potential." A step-wise approach to expansions would seem to be the simplest approach: first obtain a commitment on a single-Train Expansion and then follow that up with a case for a subsequent Expansion.

Even if the reserve base and obligations upon the producers are sufficient to accommodate a 2-Train Expansion, there are differing views as to whether it is desirable for Trinidad and Tobago to export more gas as LNG. Train 1 was market driven whereas the proposed Expansion is supply driven with lower international gas prices. Thus, two additional Trains of LNG heightens the debate on the benefits to T&T of more LNG in respect of other uses of gas that are likely to be more relevant in the longer term.

It would help the sponsors of the Expansion to counteract any negative image of additional LNG with tangible benefits in areas where the Government has indicated that it wishes to place strategic emphasis.

One obvious area is the ability of the LNG Expansion to provide volumes of ethane at prices that will facilitate an ethylene project. The obvious corollary is that if ALNG cannot supply ethane at the requisite volumes and prices, the Expansion cannot be said to be a benefit as regards an ethylene project. And, based upon proposals by the project sponsors, it is not apparent that they have appreciated the enormity (or sensitivity) of the problem.

Another area is gas-pricing terms between NGC and producers, which can then be reflected between NGC and petrochemical producers. More equitable risk sharing across the gas chain will help these gas-based industries to survive in the very competitive international climate. Security of gas supply will be critical for an ethylene project and the operations of gas-based industries contribute directly to the demand for gas and, indirectly, to the demand for ethane. I think that there is now enough history to show that cooperation is desirable particularly if time is of the essence."

After the internal disagreements among the Atlantic 1 Equity holders were resolved, they unanimously decided to pursue a 2-Train expansion. Consequently, my role as the Project Manager in Atlantic 1 under the Cost Sharing Agreement became redundant, and I left Atlantic 1 when my contract as the Train 2 Project Manager expired at the end of December 1998.

The Train 2/3 Approval. In February 2000, the Minister of Energy and Energy Industries, Senator Finbar Gangar, and NGC's chairman, Steve Ferguson, became "bitterly divided" on the issue of whether the second and third phases of the LNG project were in the Nation's best interests.[18]

---

[18] This development might have arisen as a result of the unresolved position on gas supply security in the 1996 Gas Supply Contract between Amoco and NGC and the recommendation by NGC's management, when it sought Board approval of that contract, to defer the issue to "another forum" when the Government's approval of the Atlantic LNG Train 2 approval was sought. It was the strategy that had been adopted in respect of the Train 1 project (which led to the State obtaining concessions from the shareholders in Atlantic LNG.

NGC's chairman launched a bitter attack on the deal in a presentation to a high-level meeting of the Boards of four State companies – Trinidad and Tobago National Petroleum Marketing Company Limited, The Petroleum Company of Trinidad and Tobago Limited, The Trinidad and Tobago Electricity Commission, and NGC. The security of the natural gas supply was one of the main reasons why NGC was against the terms of the LNG Expansion. NGC had argued that -

> *"The expansion would use about 50 percent of the country's proven reserves and would put domestic industries and T&TEC at risk of running out of fuel if no new gas was found and delivered"*.

Earlier, it was also reported, in the *Sunday Guardian* newspaper on 6th February 2000, that David Wight, bpAmoco's Chief Executive Officer, in a "last ditch offer to save the LNG Project," amended his proposals to the Government in two areas: security of natural gas supply; and issues about ethane.

However, David Wight's assurances concerning the security of natural gas supply applied only in respect of some of T&TEC's customers and, as indicated below, the assurances were conditional:

- *"Assurances by BP to deliver gas for critical domestic gas markets giving 'highest priority' to gas requirements for T&TEC's residential, commercial and light industrial consumers.*

- *BP Amoco will try to meet shortfalls in T&TEC's gas requirements to produce electricity if other suppliers fail to do so.*

- *If BP Amoco fails to meet the requirements of the National Gas Company, it will pay liquidated damages.*

- *In return, BP Amoco wants new tax incentives and an undertaking by the Government not to take future measures that may affect the company's earnings from exported gas."*

On Monday, 13th March 2000, the following agreements were signed in a private ceremony:[19]

(a) A Heads of Agreement permitting the expansion of the LNG facility, which later resulted in an Expansion Agreement between the Government and the Atlantic LNG 2/3 Company;

(b) A Heads of Terms for an Expansion Agreement between the Government and bp Amoco, which made provision for security of supply for natural gas and a floor price; and

(c) An agreement reducing Amoco's price of gas to NGC by 12%.

My Appointment at BG. I joined BG in Q4 2000 as the project manager/LNG coordinator for Train 4. At BG, my duties included overall responsibility for the major Train 4 commercial agreements within the Atlantic LNG forum, namely, the gas transportation agreement, the gas processing agreement, project financing negotiations,[20] and the Train 4 shareholders' agreement. I also represented BG as an Alternate Member on the Boards of the Atlantic LNG entities for the Trains 1, 2/3, and 4 Projects. Accordingly, I became intimately familiar with the commercial terms that had been agreed for the Train 2/3 Expansion, although I was not involved in those negotiations.

---

[19] 'LNG Deal Signed at Last', *The Trinidad Guardian*, Tuesday 14th March 2000.
[20] A Term Sheet was agreed with ABN AMRO but the parties eventually decided against project financing.

The Train 2/3 Expansion. The nominal capacity for Train 2 and for Train 3 was 3.3 million MTPY of LNG, and each Train utilised approximately 500 MMscfd of natural gas. The trains had identical designs, footprints, and LNG output capacities. Under the gas supply arrangements, the Atlantic LNG 2/3 company bought 50% of the required gas for Train 2 from British Gas and its Partners in the NCMA [21] and 50% of the required gas for Train 3 from British Gas and its Partners in the ECMA. [22] bpTT sold 50% of the gas to the Atlantic LNG 2/3 company for both Trains 2 and 3. As regards the LNG offtake, the Atlantic LNG 2/3 company sold LNG on an FOB basis under 20-year contracts to the gas producers (or their affiliates) at a delivery point in Trinidad. Train 2 delivered its first cargo in August 2002, and Train 3 in May 2003.

---

[21] The partners at the time were BG (45.9%), Petrotrin (19.5%), PetroCanada (17.3%) and Agip (17.3%).

[22] The partners at the time were BG (50%) and ChevronTexaco (50%).

Atlantic 1 Restructuring. In July 2000, Cabot sold its interest in Train 1 to Tractebel, and thereafter, the five participants in Train 1 (the "Atlantic 1 Equityholders"), together with Atlantic 1, restructured the original Shareholders' Agreement dated 20th July 1995. As a result of the restructuring, the Parties terminated the original Shareholders' Agreement and formed a new Delaware limited liability company, Atlantic 1 LLC Holdings ("Atlantic 1 LLC"), which owned 100% equity interests in Atlantic 1. The directors from the five previous shareholders became member representatives of Atlantic 1 LLC, the offshore LLC Company.

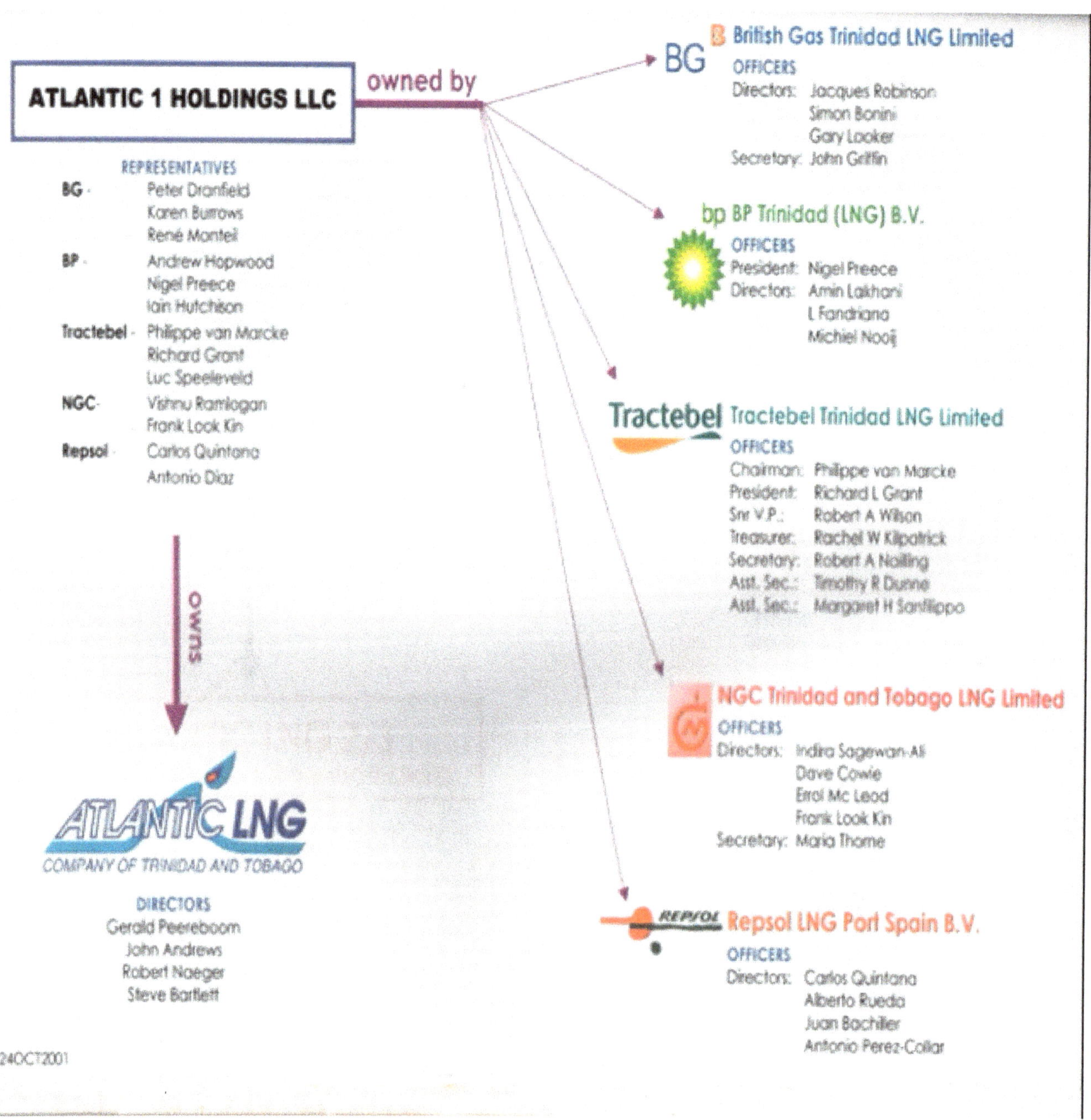

The Joint Use and Operating Agreement ("JUOA"). The restructuring of Atlantic 1 was driven by two motives: firstly, to avoid the shareholders owing any fiduciary or corresponding duties to Atlantic 1, and secondly, to get around the rights granted to minority shareholders under the law in Trinidad and Tobago. [23]

As part of the restructuring, it was contemplated that additional natural gas liquefaction units would be constructed, owned, and operated in accordance with the JUOA. The JUOA also imposed a requirement on investors in any Expansion to adopt a holding LLC company structure and to organise the proposed Train Owner as an unlimited liability company under the Companies Act, with articles and by-laws in the same form as the articles and by-laws of Atlantic 1 TT, the local entity.

---

[23] The Act "provides a new form of protection to minority shareholders who disagree with a particular fundamental change. This takes the form of a right to dissent from such change and to have the company purchase their shares at their fair value. The impact of the new right to dissent is potentially significant. A company faced with paying the fair value to a large number of dissenting shareholders could face liquidity problems. In practical terms, therefore, the existence of the right to dissent may in some circumstances inhibit the majority from making fundamental changes which they might otherwise have pursued". Hamel-Smith, *'Fundamental Changes and Minority Protection'*, http://trinidadlaw.com/fundamental-changes-and-minority-protection/

Under the Atlantic 1 LLC Agreement and the JUOA, the parties agreed that the purpose and business of Atlantic 1 TT was to own and operate Train 1 and to manage and operate all LNG trains pursuant to the JUOA. The Atlantic 1 LLC Company was deemed to be the offshore decision-making body in major matters, leaving Atlantic 1 TT to implement and manage the day-to-day business. Accordingly, after the restructuring, Atlantic 1 TT had two interests: firstly, as the operator of all LNG Trains ("the Operator"); secondly, as a wholly-owned subsidiary of Atlantic 1 LLC, the owner of Train 1. Ultimately, Atlantic 1 TT managed the business and affairs of the company through a process that involved shareholder recommendations, approval, and sanction via the Members' representatives in the offshore entities. The Figure below illustrates the re-organised structures for the three LNG Projects pursuant to the JUOA.

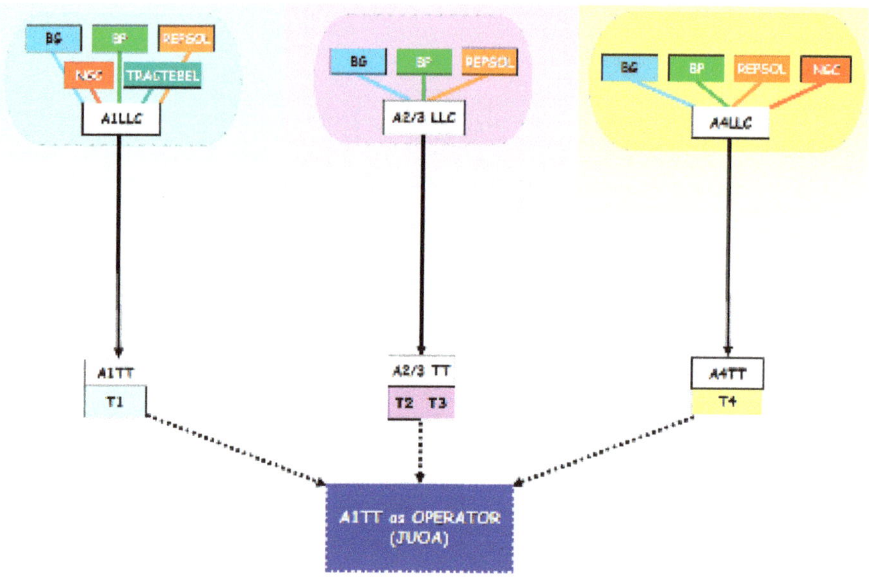

As the Operator, Atlantic 1 TT was required to manage, operate, and maintain the LNG Facilities at Point Fortin:

(a) in accordance with the standard of a Reasonable and Prudent Operator[24] and

(b) in a manner that allowed Atlantic 1 TT to perform its obligations under each of the Train 1 Project Agreements.

---

[24] The "reasonable and prudent operator" (RPO) standard, which commonly arises in oil and gas contracts, is a standard against which to measure the performance of primary obligations. In *Scottish Power UK Plc v BP Exploration Operating Company Ltd*), [*2015*] *EWHC 2658* (Comm) (25 September 2015), the court held that the standard should be interpreted by giving the natural and ordinary meaning of the obligation to perform a contract.

Commercial Structures. Although the structure of the Train 2/3 project legally reflects a buy/sell arrangement, in practice, it was a *de facto* tolling arrangement.

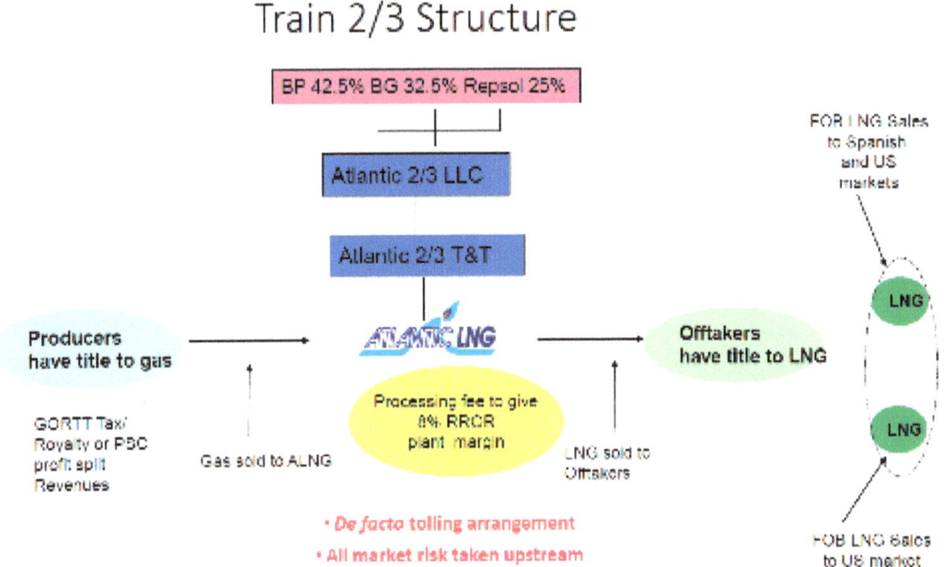

Under the Train 2/3 commercial arrangements –

(a) The gas producers sold the gas to the Atlantic 2/3 TT company.

(b) The Atlantic 2/3 TT company sold the LNG to the gas producers or their affiliates.

(c) The Atlantic 2/3 company obtained a guaranteed utility RROR of 8%.

(d) The affiliates of the gas producers marketed the LNG.

(e) All market risk was taken by the upstream companies.

Thus, the Train 2/3 commercial structure sharply contrasted with the original Train 1 merchant structure, which had the following key commercial elements:

(a) Atlantic 1 TT purchased the natural gas feedstock from Amoco and produced LNG.

(b) Atlantic 1 TT sold the LNG (which it owned) to Cabot and Enagas.

(c) Atlantic 1 TT took the market risk in the pricing arrangements under the LNG Offtake Agreements with Cabot and Enagas.

(d) The price of the gas purchased from Amoco was linked to the market price of LNG, so the price risk was shared between the upstream and Atlantic 1 TT.

Train 4. At a meeting on 9th September 2003, an Expansion Resolution was approved for Train 4 using, for the first time, the procedure set out in the JUOA.

Expansion Procedure

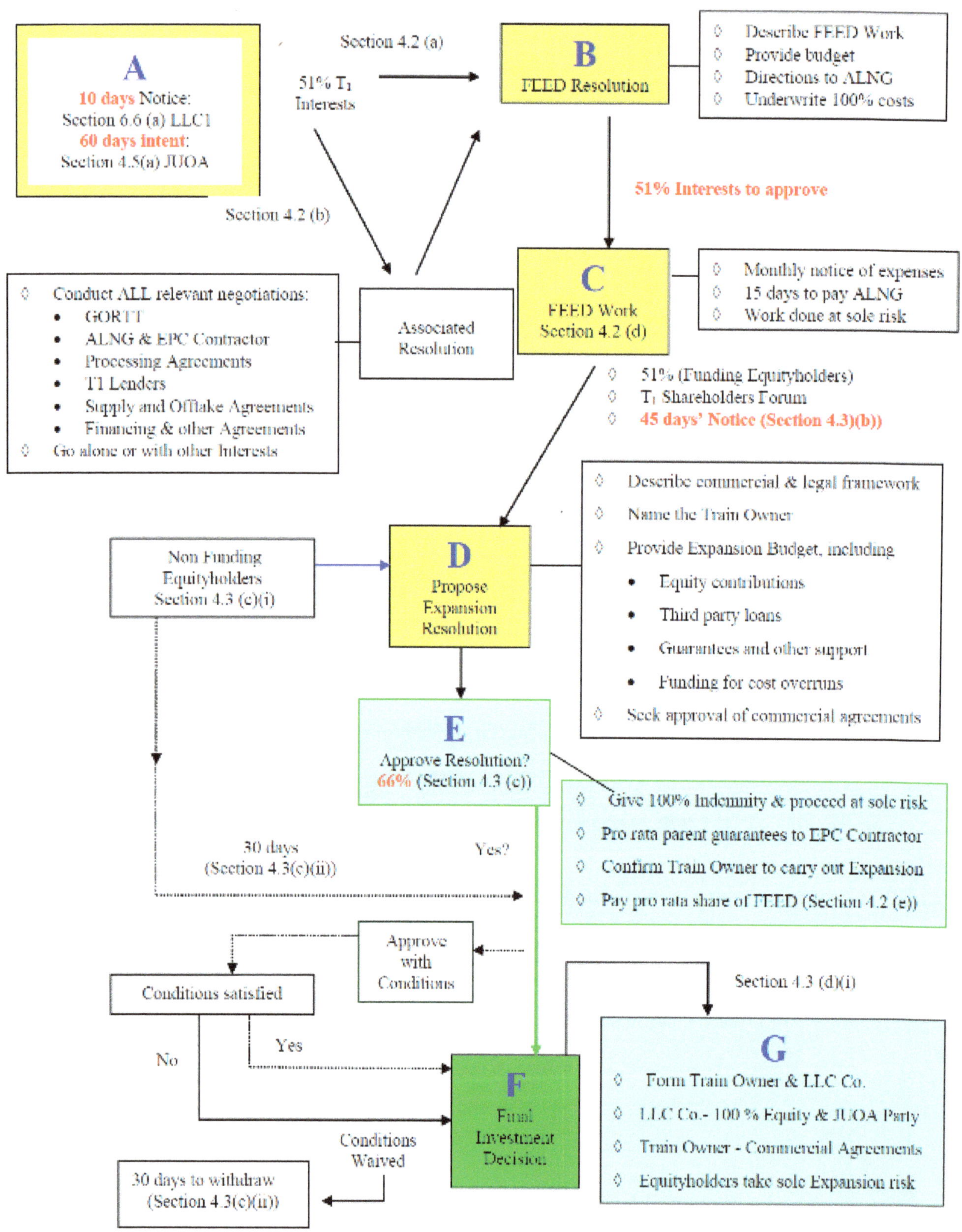

The commercial structure for Train 4 reflected a pure tolling arrangement in which each sponsor was able to obtain LNG volumes in proportion to its gas supply. With Train 4 –

(a) The gas producers (or their affiliates) had legal title to the LNG that was produced.

(b) The LNG marketing was carried out by subsidiaries of the gas producers (not Atlantic 1).

(c) The value capture for the LNG processing entity was a prescribed 8% rate of return.

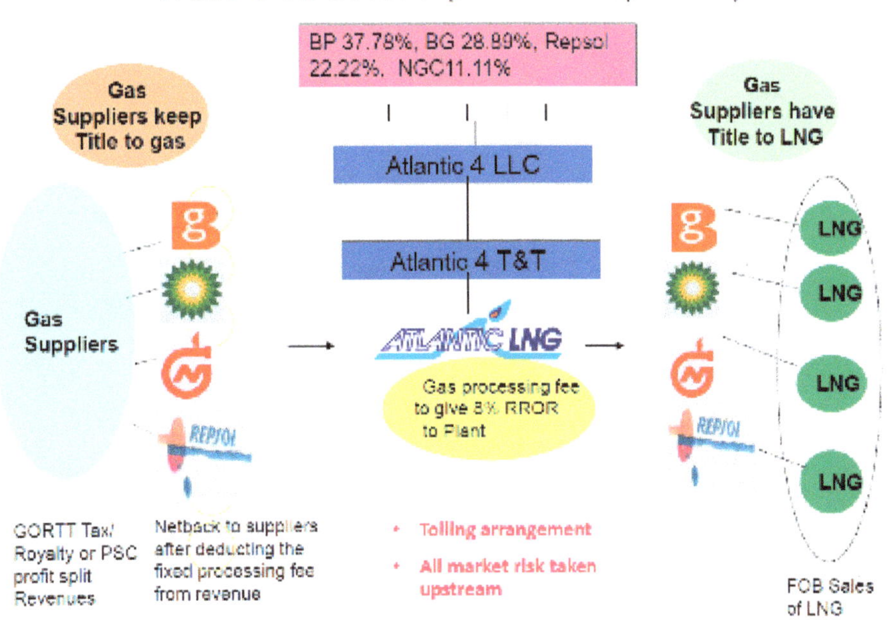

Thus, the Train 2/3 and Train 4 commercial structures were conceptually very similar tolling arrangements but both sharply contrasted with the Train 1 merchant structure.

Train 1 Structure

Notably, for Train 4, the Government did not have audit rights in respect of **all** aspects of the LNG business of the marketing affiliates of the gas producers, which would have exacerbated the challenges of determining the appropriate netbacks to the wellhead when the gas producers (or their affiliates) completely controlled LNG marketing. This was one of the consequences of the tolling structure, which allowed the gas producers to bypass Atlantic 1, in which NGC had an interest. In 1998, before the Government approved the Train 2/3 *de facto* tolling structure, I had stressed (among other things) the strategic role of Atlantic 1, which had provided an intermediate institutional framework to prevent complete vertical integration by the gas producers. In that regard, Atlantic 1, like NGC's ownership of the onshore pipeline transportation system, had played a strategic role.

The World's Largest LNG Train. Train 4 had unique project development features. At the time it was constructed, with a capacity of 5.2 million tonnes of LNG per annum, it was the world's largest LNG train.

The 56″ Pipeline. The project required the development of a new Cross-Island pipeline system to transport gas from the gas aggregation facilities at Beachfield to the Atlantic LNG site. Ultimately, the Train 4 sponsors chose a 56″ pipeline, which was built by NGC with assistance from bp. This pipeline was one of the largest diameter pipelines that had ever been built. After commercial negotiations, the parties agreed that the new 56″ pipeline would be linked with the 36″ pipeline that supplied Trains 1-3 and that NGC would operate all aspects of the unified system.

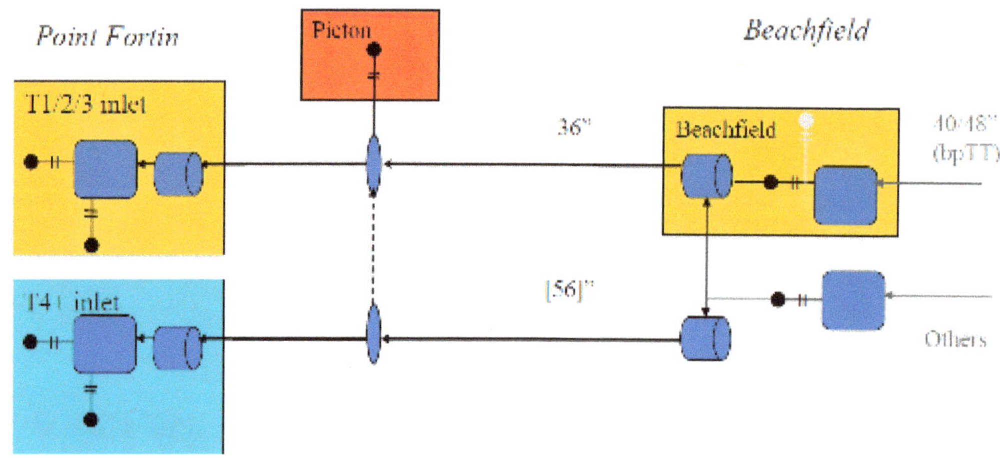

The 4th Tank and 2nd Jetty. It was agreed to include a 4th tank based upon a Lloyd's Maritime recommendation, although the matter was hotly debated among the proposed investors in Train 4. It was also agreed to include a new jetty. The new jetty, with the same loading capacity of 10,000 m³/hr. as the first jetty, was added in order to allow the simultaneous loading of two LNG ships.

Noise. At the time, although it was unclear whether the site had ever been classified as an Industrial Site for the purposes of the Trinidad & Tobago rules, it was recognised that the site had been used for industrial development since the early 1900s when UBOT commenced operations. As part of project development studies for Train 4, Atlantic 1 TT procured the services of specialist noise consultants to carry out field measurements and modelling in order to evaluate the noise levels at the proposed site based upon National and World Bank standards.[25] Based on Train 1 noise levels, modeling work was done in order to factor in the noise levels contributed by Trains 2, 3, and 4.

Field measurements of sound levels were completed both during normal plant operation and during a Train 1 shutdown. The results (measured in leq dBA) refer to the five zones below.

	ZONE 1 (NE ZONE)	ZONE 2 (EAST OF PLANT)		ZONE 3 (MELVILLE AREA)	ZONE 4 (S. BEACH AREA)	ZONE 5 (HOSPITAL AREA)
		Trinmar	*Housing*			
Background	48	43-51	46-52	46-53	51	42-50
Train 1	55	53	58	52	53	47
Trains 1, 2	55	54	59	52	55	46
Trains 1,2,3	56	56	61	55	58	45
Trains 1,2,3,4	55	58	62	59	62	45

The measurements confirmed that Train 4 had complied with the Trinidad and Tobago Noise Rules as well as the more stringent World Bank Rules.

The Gas Processing Agreement. Under the JUOA, the processing tariff for Train 4 had to be –

---

[25] The relevant World Bank standards provide for a maximum of 70 dBA but also permit an increase in the existing ambient level of **3 dBA** above background. Under the 2001 Trinidad and Tobago Rules, there is **an upper limit of 75 dBA** in respect of an Industrial area, such as the site at Point Fortin, and an upper limit of 65 dBA for residential areas.

(a) *"similar in structure to the plant margin for the Train 2/3 Expansion"* and

(b) *"result in a rate of return and risk profile to the Train owner similar to that of Atlantic 2/3."*

For Train 4, the Parties had used a hybrid of the Train 2/3 gas supply and LNG offtake contracts (with appropriate modifications) to produce the gas processing agreement. Thus, the focus was changed from LNG offtake (as per the LNG 2/3 Expansion) to the gas supply that would be necessary to provide the desired quantity of LNG. As a result of this change in focus, many commercial issues arose, mainly between BG and bp, but the two most contentious were:

(a) *Capacity utilisation.* Should the gas supplier be obliged to give its written approval before any other supplier could utilise its gas inlet entitlement to processing capacity that was not being fully utilised? BG argued that since the gas inlet entitlement had been paid for, each person should be required to negotiate with the other Parties in respect of processing rights (or at least compensate a person at the agreed plant margin). The concern was that once a person gave up control by allowing Atlantic 1 TT to utilise processing capacity based on gas supply availability, it potentially provided an advantage to bp to capture processing capacity on a day-to-day basis because of bp's more secure gas supply position.

(b) *The NGL Threshold for allocating NGLs.* In the Train 2/3 agreements, under certain circumstances, a gas supplier whose gas was at or below the NGL Threshold for the allocation of NGLs, would **not** be allocated NGLs during the particular month. The difficulty for BG in Train 4 was that gas from the NCMA (with a high percentage of methane) could be distinguished from BG's supplies from the ECMA. If the gas from the ECMA on its own would be over the NGL Threshold, but the effect of commingling with the gas from NCMA gas would be to take the commingled stream below the Threshold, the ECMA partners might claim that they were losing value.

After the resolution of the issues, the parties approved an Expansion Resolution at a meeting of the Atlantic 1 Equityholders on 9th September 2003, and Train 4 began its commercial operations on 1st May 2007.

Chapter 6: The Potential of an Aluminium Industry

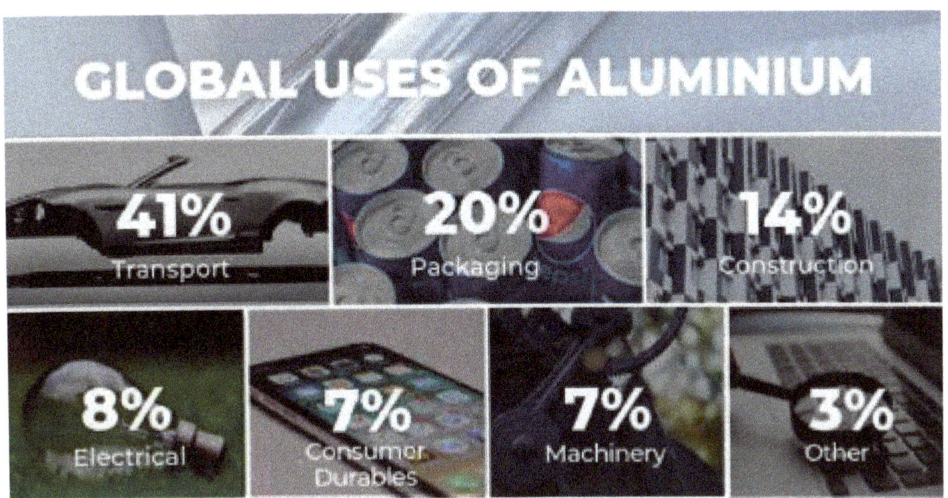

Aluminium Occurrence. Aluminium,[1] the chemical symbol Al, the most abundant metallic element in Earth's crust, occurs chiefly in bauxite (its principal ore). Bauxite is a reddish clay-like substance that is an impure mixture of earthy hydrous Aluminium oxides and hydroxides.

---

[1] *Aluminum is preferred in North America,* whereas Aluminium is the spelling preferred in the United Kingdom and most other English-speaking nations, https://www.merriam-webster.com/words-at-play/aluminum-vs-aluminium.

Bauxite is converted into alumina (Aluminium oxide, Al_2O_3), a white substance, which is used as a starting material during the process for the manufacture of Aluminium metal.[2]

Properties. Aluminium and its alloys have many properties that make them useful in a wide range of applications across an impressive range of industries.

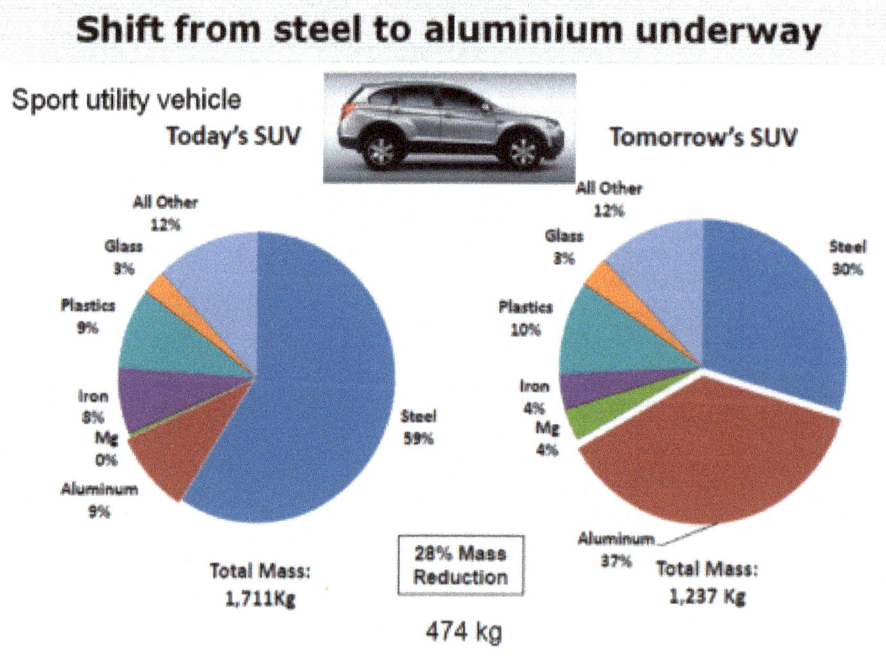

---

[2] As a general rule, 4 tons of dried bauxite is required to produce 2 tons of alumina, which, in turn, produces 1 ton of Aluminium. (International Aluminium Institute, *Alumina Production*, 26th January 2022, https://international-Aluminium.org/statistics/alumina-production/).

Manufacture. The Hall-Héroult process is universally used to produce Aluminium. It is an electrolytic process in which Aluminium oxide is dissolved in molten cryolite (Na_3AlF_6), the electrolyte, and then reduced electrolytically to Aluminium at a temperature of around 960°C. The smelting takes place in a steel vat called a reduction pot. The bottom of the pot is lined with carbon, which acts as one electrode of the system. The opposite electrodes consist of a set of carbon rods suspended above the pot. During the electrolytic process, alumina is reduced to Aluminium that accumulates on the floor of the pot, and carbon reacts with oxygen to form carbon dioxide. Reduction pots are arranged in rows (called "potlines") consisting of 50-200 pots that are connected in series to form an electric circuit.

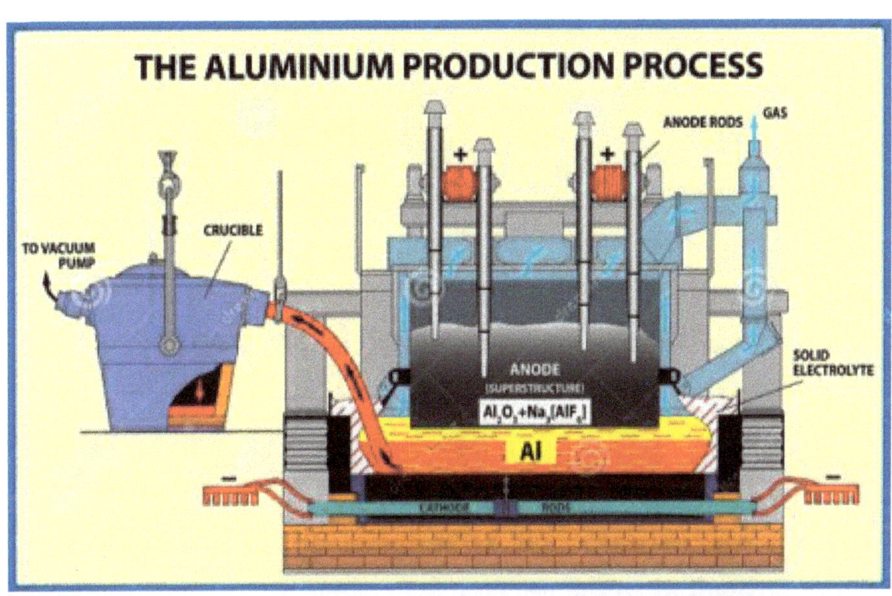

Technology. Two main technologies are used in smelter plants throughout the world – the Söderberg Cell technology and the Pre-bake Cell technology. The older Söderberg technology uses a continuous anode which is delivered to the pot in the form of a paste, which bakes itself. The alternative, the Pre-bake Cell technology, uses multiple anodes in each cell which are baked in a separate facility and attached to 'rods' that suspend anodes in the cell. New anodes are exchanged for spent anodes, which are then recycled into new anodes.

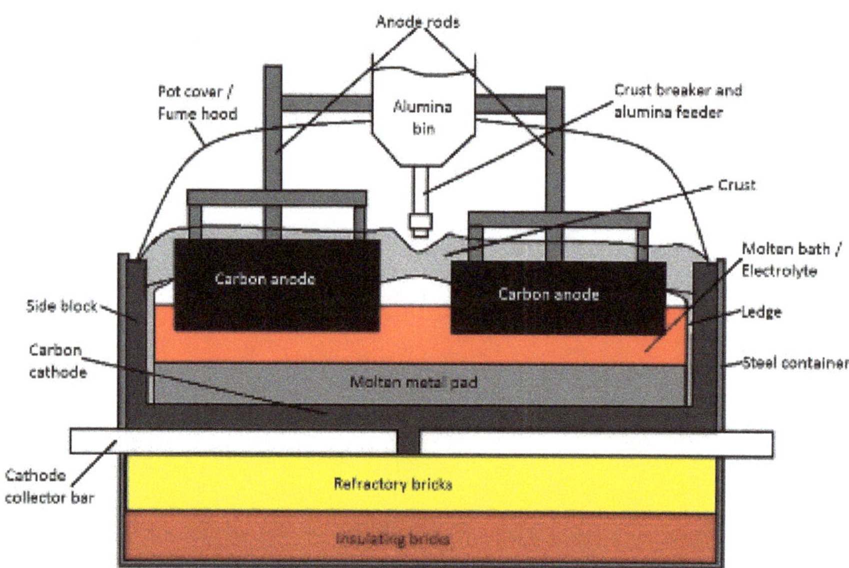

Health &Technology. Compared with the more modern Prebaked anode technology [3]–

   (a) Soderberg cells are considered to be less efficient and have higher production costs;

   (b) They are more difficult to automate; and

   (c) They present the greatest environmental and health challenges.

Today there are a few Soderberg smelters operating in North and South America with a capacity that has been estimated to be less than 1 million tonnes per year. On Friday, 13th March 2009, at 09.01 a.m., the power in the Søderberg potroom at the Aluminium plant at Karmøy in Norway was switched off, marking the end of production in the very last electrolytic cell in Hydro, the large industrial company, to use the old Søderberg technology.[4]

---

[3] Mike Barber and Alton Tabereaux, 'The Evolution of Søderberg Aluminum Cell Technology in North and South America', Journal of the Minerals, Metals & Materials Society 66(2), December 2013.
[4] Hydro, 'Last day for Søderberg technology in Hydro', https://www.hydro.com/en/media/news/2009/last-day-for-soderberg-technology-in-hydro/.

High/low-Cost Producers. The U.S., Western Europe (except Norway and Iceland), and Japan are among the high-energy-cost Aluminium producers, while typical low-cost producers are Canada and Brazil (Hydroelectricity), Australia (coal), and some Middle Eastern countries (natural gas). Under all energy supply scenarios, an Aluminium smelting project will benefit by having lower production costs if there is a power supplier who can provide reliable and safe power to both the smelter and the local community.

Associated Manufacturing Processes. There are many processes associated with the Aluminium industry. They can be classified in three areas, namely, the primary industry, which produces the metal; the secondary industry involving extrusion, casting, and rolling; and the recycling industry.

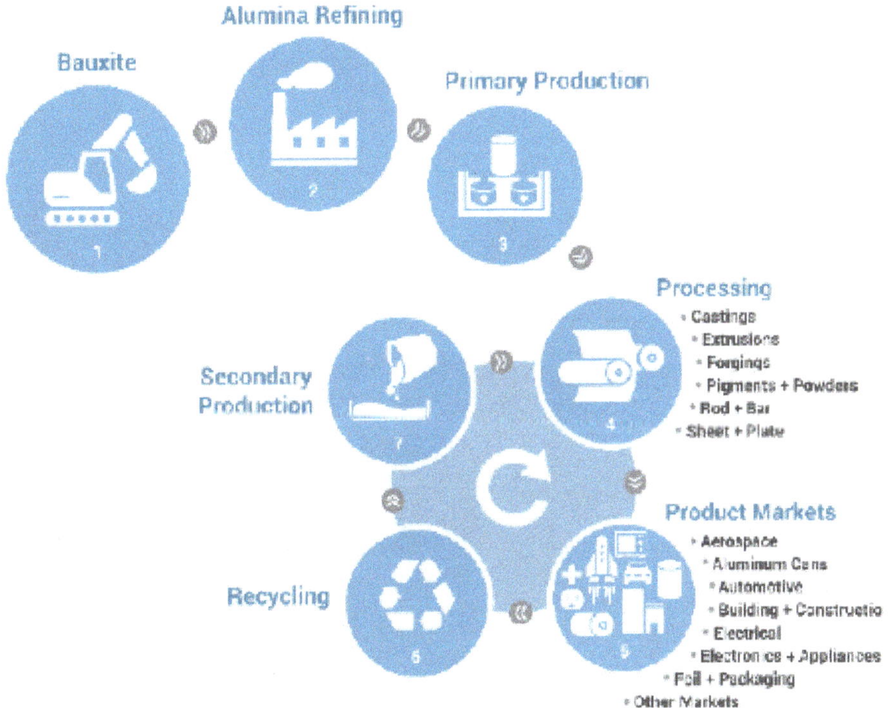

Trading. The London Metal Exchange ("LME"), one of the oldest commodities exchanges in the world, was established in 1877. In 1978, the exchange trade for Aluminium contracts started on the LME, and since then, the price for primary Aluminium has become uniform all over the world. The LME prices are considered to be the standard global prices for base metals, which include Aluminium, Zinc, Lead, Copper, and Nickel.

Energy Consumption. All steps in the Aluminium production process, as with all industrial processes, consume energy but the energy required by these processes is relatively low compared to the electrical energy required by the reduction process to convert alumina to Aluminium. Low-cost power is the main determinant for the economic viability of Aluminium production, and the availability of cheap electric power has led to shifts in the location of smelters.

Projects in Trinidad and Tobago. During my tenure as the Vice President, Business Development, NGC had been involved in discussions with various participants to establish an Aluminium smelter in Trinidad and Tobago. Later, briefly, as a member of the Board of Alutrint Limited ("Alutrint"), I was involved with the Alutrint project. The imperatives driving all of these initiatives had been the availability of the relevant raw materials for Aluminium smelting in the region and natural gas for power generation in Trinidad.

The Norsk Hydro Project. At NGC, the most advanced project occurred with Norsk Hydro, a member of the Norsk Hydro Group ("Norsk"), Norway's largest publicly owned industrial concern. Norsk Hydro ASA was one of the largest Aluminium companies worldwide, with operations in some 50 countries around the world, and was active on all continents. The salient features of the Norsk Project were:

(a) In 1997, Norsk had forecasted an unsatisfied demand of primary Aluminium of 3.5 million tonnes of Aluminium by the year 2004, primarily in Europe and the Far East, and had proposed a smelter with a capacity of almost 200,000 MTPY costing about U.S. $1 billion.

(b) Construction costs in Trinidad and Tobago were estimated to be 14-15% lower than in Western Europe, and Norsk had completed the pre-feasibility phase for the project.

(c) Norsk had confirmed to NGC that, after the pre-feasibility stage, an Aluminium smelter project had looked attractive and that they had taken a decision to proceed to the next phase, which would involve site and soil studies and more detailed quantification of infrastructure, port, civil and environmental costs.

(d) Under the terms of a proposed site lease, Norsk had paid Plipdeco and had secured an option to use a site for the project (adjacent to the Hydro Agri site at Point Lisas).[5]

(e) It was contemplated that in the first instance, about 500,000 tons of alumina raw material would be shipped to Trinidad, and 200,000 MTPY products would be exported, possibly expanding to 400,000 MTPY with a 2nd pot line that utilised common facilities.

Security of Power Supply. A reliable and uninterrupted power supply (with the attendant gas supply) to prevent unplanned shut-downs is a critical feature of a smelter operation. Power failures can occur without warning, and a power failure of more than a few hours can severely damage prebaked anode potlines.[6] Potline power interruptions of 10-30 minutes are commonly carried out in many smelters to change cathodes and perform other necessary repairs. A prebaked cell can normally tolerate short power interruptions without too many adverse effects. However, there can

---

[5] Another project, the Kaiser group project, had held discussions with Plipdeco and had proposed to locate its project on a site, north of the Couva river, which had not yet been acquired by Plipdeco from Caroni Trinidad Limited.

[6] Harald A Øye and Morten Sørlie, 'Power failure, restart and repair', *Aluminium Today*, https://aluminiumtoday.com/content-images/news/Oyeweb.pdf.

be significant operational side effects of a full power interruption for five or more hours. A smelter project, therefore, often has its own power island connected to the national grid.

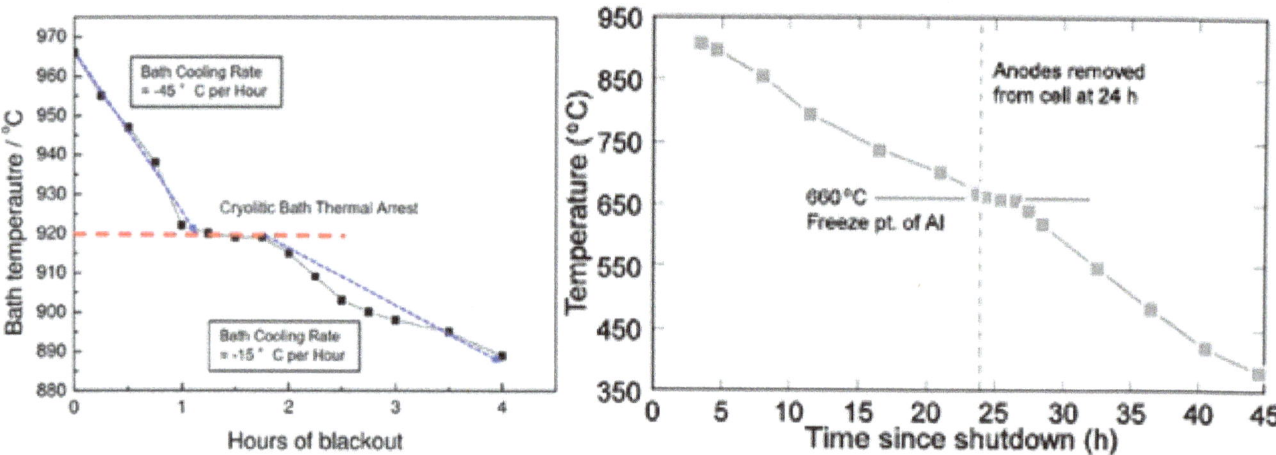

Natural Gas Supply. The anticipated gas requirements were 48 MMscfd or 82 MMscfd based, respectively, on either combined or open cycle operation of the gas turbines for power generation.[7] Norsk requested an uninterruptible gas supply contract for the delivery of gas at an hourly rate of 2.5 or 3.0 MMscf at a delivery pressure of 580 psi (40 Bar, 4000 KPa).

---

[7] A combined-cycle power system typically uses a gas turbine to drive an electrical generator, and recovers waste heat from the turbine exhaust to generate steam. The steam from waste heat is run through a steam turbine to provide supplemental electricity. The overall electrical efficiency of a combined-cycle power system is typically in the range of 50–60% — a substantial improvement over the efficiency of a simple, open-cycle application of around 33%.

Under the terms proposed for the gas supply contract, Norsk would not accept, as an event of "*force majeure*," the unforeseen decline of gas reserves in specific fields nor the failure of NGC's suppliers to supply gas to NGC, unless such failure was itself caused by *force majeure* by the gas supplier. The proposed gas pricing model for the Project indexed a gas price to the LME price for Aluminium as well as the average price of coal to the electric utilities in the United States.

Electricity Supply. Under the terms of a Memorandum of Understanding with NGC, Powergen and T&TEC agreed that the parties should work together to develop an energy proposal that would maximise the reliability of the energy supply and minimise the costs of energy to the Aluminium smelter.

Government Approval of the Norsk Project. On 12th November 1998, former Prime Minister Basdeo Panday delivered the feature address at the signing ceremony for a project agreement to establish an Aluminium smelter proposed by Norsk. Mr. Panday's remarks provide important background for (and the benefits of) developing an Aluminium industry in Trinidad and Tobago.

"It is with great pleasure that I address you today on what, for my Government and the country, is a most important and an historic occasion. Today heralds a new dawn - the start of what might be considered a new industry with the signing of the Project Agreement for the establishment of an Aluminium Smelter in Trinidad and Tobago.

The road to the establishment of an Aluminium Smelter in this country has been by no means an easy one. The earliest initiatives can be traced back to the late 1960s following the discovery of significant natural gas reserves off the south-east coast of Trinidad. Since then, there have been at least three major sets of initiatives, both by government and by the private sector -all of which have been unsuccessful.

In 1974, there were discussions between the Governments of Trinidad and Tobago, Jamaica, and Guyana to establish a regional Aluminium Smelter. The Regional Smelter Initiative was actually initiated by Guyana in a plan to develop integrated hydro/smelter facilities. A proposal was developed to draw on the raw material from Guyana and energy from Trinidad in the first phase. Technical Studies indicated that the economics favoured a plant of 150,000 metric tonnes per year and the viability of the project had a high sensitivity to the c o s t of electricity and to the cost of alumina. In 1977, the Government of Trinidad and Tobago took the fundamental decision comprising Kaiser Engineers, GEC, and PCI.

However, all of these initiatives failed to 'materialize.

Honoured and distinguished guests, there is a saying that "The only real mistake is the one from which we learn nothing." Well, this Government has learned from the mistakes of the Past. Indeed, we have excelled in areas in which previous attempts to achieve what is now a reality have failed miserably.

There are important socio-economic bases for the establishment of a world-scale Aluminium Smelter in Trinidad and Tobago. There are abundant supplies of some of the world's best quality bauxite deposits in our sister CARICOM nations - Jamaica and Guyana. This factor, coupled with the fact that Trinidad and Tobago is blessed with adequate supplies of competitively-priced natural gas-based energy from experienced and proven gas supply companies, a well-developed natural gas infrastructure, and an experienced workforce. And when these factors are combined with the other enabling factors, such as the hospitable investment climate, a stable government, and political system, and the favourable geographical location of the country at the gateway to potential markets in the Americas and Europe, it is clear why Norsk Hydro has selected Trinidad and Tobago as the location for its next major investment in Aluminium Smelter facilities.

The Global Outlook for the Aluminium Industry in terms of demand forecast is encouraging. Projections are that while in 1997, primary Aluminium consumption in the Western World was 18.6 million tonnes, estimates are that by the year 2006, consumption will reach 23.5 million tonnes resulting in an annual growth rate of 3%. On the demand side, projections are that by the year 2006, there will be 5 million tonnes of new demand requiring the establishment of two new smelters per year.

Trinidad and Tobago's thirty-year quest to establish a local smelter is finally within our grasp as we sign this Project Agreement for the establishment of an Aluminium smelter/power plant. This project will provide sustainable employment opportunities and further aid economic development in this ·country. It is expected that the plant will directly employ over 550persons in permanent jobs. During the construction period, it is estimated that over 2500 jobs will be created. Relative to other energy industries in the natural gas-basedsector, an Aluminium project would generate substantially more highly skilled jobs, and it promises to create significant downstream opportunities.

Moreover, apart from its employment-generating potential, there is, of course, the macro-economic impact to be derived from Petroleum Profits taxes and royalties from both the natural gas and condensate produced, corporation taxes, foreign exchange earnings, and from the downstream opportunities for the local manufacturing sector. An Aluminium smelter would also provide the opportunityfor a significant diversification of the industrial base of the country.

The project agreement makes provision for an undertaking by Hydro to supply Aluminium to local and other investors for the manufacture of semi-fabricated and other products from Aluminium. It is envisaged that there will be business opportunities for manufacturers engaged in producing semi-fabricated products such as window and door frames and in Aluminium castings and forgings.

At this juncture, I reiterate that an investment in the establishment of an Aluminium smelter and related facilities in Trinidad and Tobago will be the country's, and possiblythe region's, single largest, capital investment. Over a ten-year period, it is estimated that direct investmentwill be in the order of U.S.$2.6 billion for the two-train smelter and power plants. In addition to the multiplier effects in terms of indirect employment creation, this investment would facilitate the expansion eastwards of the Point Lisas Industrial Estate, an expansion of port and marine infrastructure, and the increased sale of natural gas.

Another significant benefit of this smelter investment is the long-term commitment to the economy of Trinidad and Tobago, with plant life exceeding 40 years. This will, of course, require a long-term demand for new skills and human resource training. A new era is dawning before our very eyes. ... .

Trinidad·and Tobago is now a prime location for foreign investment, placing us amongst a select group of developing countries. A new outward-looking approach has resulted in levels of investment success that can, in part, be attributable to the Government's focus on private sector-led development in concert with our aggressive moves to monetize current gasassets.

By the year 2000, Trinidad and Tobago will be the locationfor an enviable array of oil, gas, and downstream concerns. There is now an air of cautious optimism that the range and intensity of activity in both the upstream and downstream sectors of the Trinidad and Tobago energy industry will bode well for the continued success of the country's industrialization thrust in a now relentlessly globalized world economy.

Our prudent social and economic policies combined with good economic results have gained recognition with the international financial community. The positive performance of our local economy and our now sound financial standing have made us quite attractive as an investment location. All these favourable conditions and positive economic developments have not gone unrecognized. Multinational companies and other investors have already made their move to invest in Trinidad and Tobago and to take advantage of our good access to international markets.

"Well done is better than well said."

These words of Benjamin Franklin have echoed through my thoughts for some time. I feel that all of you here will agree with me when I say that we have moved a significant distance towards this ideal.

The Government is committed to attracting even further levels of investment, both local and foreign. In this scenario, the role of Government is to be the facilitator, providing attractive investment incentives along with proper infrastructural support for the conduct of economic endeavours.

In the agreement to be signed shortly, Hydro has agreed that they will maximize the use of local content in the engineering, construction, and procurement for this project. This will ensure that this industrial facility will be constructed and operated within acceptable Environmental standards. In addition to these broader benefits to the national economy, there are benefits to the other stakeholders in this project. These include the National Gas Company, Caroni(1975) Limited, T&TEC, and Amoco Power Resources Limited.

For the National Gas Company, the successful conclusion of negotiations for the investment of an Aluminium project represents a key element in the diversification of its gas sales portfolio. Investment in the project means, for NGC, additional natural gas sales in the order of 200 million standard cubic feet of gas per day for a minimum term of 25 years. For Caroni (1975) Limited, this project will mean the transformation of low-value agricultural lands into a higher-value industrial estate. With the requirement for skilled workers in the Aluminium smelter and the proposed extensive training commitment, abundant opportunities will be available for retraining of workers and the development of new skills. For T&TEC, there would be the opportunity to acquire additional bulk power from the development of new electricity generation facilities at a competitive rate. For Amoco Power Resources Limited, this project will provide an opportunity to develop a power generation facility in conjunction with Hydro, an experienced IPP - Independent Power Producer, and other entities.

For the Project Sponsor, Norsk Hydro, investment in Trinidad and Tobago was conditioned by the fact that it isno stranger to our shores. Norsk Hydro already has extensive petrochemical investments at Point Lisas in the form of sole ownership of the 250,000-tonne, Hydro Agri ammonia plant and a 49%shareholding in TRINGEN's two ammonia plants with capacities of 400,000 tonnes and 500,000 tonnes, respectively.

The coming together of these diverse concerns to bring to life this dream of an Aluminium smelter in the country is a very real manifestation of my Government's core unity message. This type of agglomeration of differing interests has become the standard by which business is being done in the country. In fact, I say that you could not have chosen a better location to bring this effort to fruition than in the twin- island, multicultural state of Trinidad and Tobago.

I am pleased to see that Norsk Hydro envisages a two-train smelter facility, each with a rated capacity of 237,000 tonnes, costing approximately U.S.$1.6 billion and requiring 400 MW of new electrical power. It is projected that the first train is scheduled for completion by the year 2002, with the second train to be constructed 5-7 years later at a cost of U.S.$1 billion. However, it is important to keep in mind what constitutes the critical success factors in business success. They are innovation, swiftness to market, and the "fit" of the business operation within the economic and infrastructural make-up of the investment site.

I would like to assure the principals of Hydro, who have taken the initiative to invest in this country, of the Government's continuing commitment to providing the best investment climate possible. An investment of this magnitude, U.S.$2.6 billion, is no doubt based on the fact that the country possesses good characteristics for success and that their principals agree with this assessment.

I wish Norsk Hydro speedy progress in the continuing site evaluation and other detailed engineering studies leading to the final investment decision for the smelter and associated facilities. I am, therefore, sure that I speak on behalf of all the citizens of this country when I say that we warmly welcome Norsk Hydro's continued confidence and commitment to the development of Trinidad and Tobago.

Ladies and gentlemen, I thank you."

Ultimately, Norsk decided not to proceed with the project because of a sharp decline in commodity prices on world markets in the wake of Asia's economic and financial crisis.

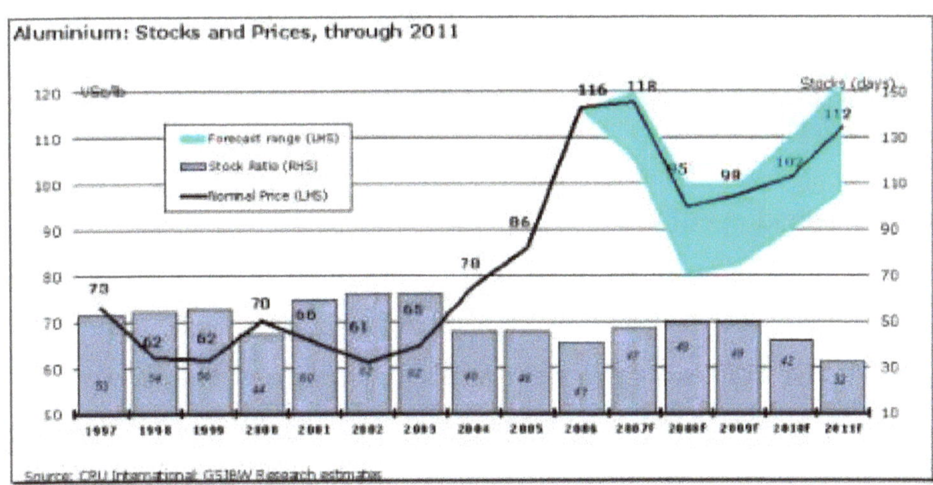

The Alutrint Project. Reuters gave the following press release confirming that the Alutrint Aluminium Project was cancelled by the UNC administration a few months after it formed the Government in 2010.

"PORT OF SPAIN, Sept 8 (Reuters)[8] *- Trinidad and Tobago's government said on Wednesday it was cancelling a $600 million project to build a 125,000 tonnes-per-year aluminum smelter in the Caribbean nation.*

> *'In addition to the health and environmental risk, there is also serious concern as to Alutrint's viability and the optimal use of our gas. This project shall cease,'*

Finance Minister Winston Dookeran said during a presentation of the 2010-2011 national budget.

Brazilian conglomerate Votorantim Group has a 40 percent stake in the proposed 125,000 metric-tonnes-per-year aluminum smelter complex, while the Trinidad and Tobago government held the remaining 60 percent. China Exim Bank was providing a credit facility of $400 million for construction of the project. The cancellation of the Alutrint smelter complex effectively ends a court battle over the project.

---

[8] Reuters, 'UPDATE 1-Trinidad cancels $600 million aluminum smelter project', September 8, 2010, https://www.reuters.com/article/trinidad-aluminum-smelter-idafn0813743920100908.

Environmental groups had openly lobbied the Government to cancel the Project, but it cannot be precluded that many people were either not sufficiently informed (or were misinformed).

Evaluating Mr. Dookeran's Comment re the "Health and Environmental Risk."

(a) **Red Mud**. After criticism in an article about mud lakes and the environmental impact of the Alutrint Project had been published in one of the daily newspapers, I explained to the author that the Project intended to import alumina, a white solid, which was manufactured from bauxite. He had assumed (wrongly) that red mud lakes would be created locally.

(b) **Pre-baked Technology**. The modern Aluminium industry had demonstrated worldwide, to have manageable environmental impacts. A major environmental concern for smelter operations was linked to the use of the older Soderberg technology in which Hydrogen Fluoride (HF) gases are generated in the electrolytic cell. However, virtually all new plants and plant expansions around the world had been based on pre-bake technology, which led to energy efficiency and encouraged recycling; therefore, waste minimisation. The Alutrint Project, like most of the world's modern smelters, had been designed to use prebaked anodes that generated less HF gases. Moreover, other gaseous effluents, polycyclic aromatic compounds, and volatile organic compounds that are generated in the manufacture of the carbon anodes would have been avoided by using the prebaked anodes.

(c) **Fluoride Emissions Control**. The emission of fluorides is one of the most debated aspects of Aluminium smelting. In the Alutrint Project, the volatilised fluorides and gaseous HF would have been collected with other gases that evolved from the pots by gas-collecting hoods. These substances would have then been passed through large ducts to central gas scrubbers, where the fluorides would have been absorbed and recovered for reuse in the pots. This system would have prevented air pollution, conserved valuable resources for recycling, and because it was a dry process, there would not have been liquid wastes for disposal. Alutrint had proposed to invest U.S.\$15.5 Million in upgrading to the most efficient fluoride scrubbing system worldwide.

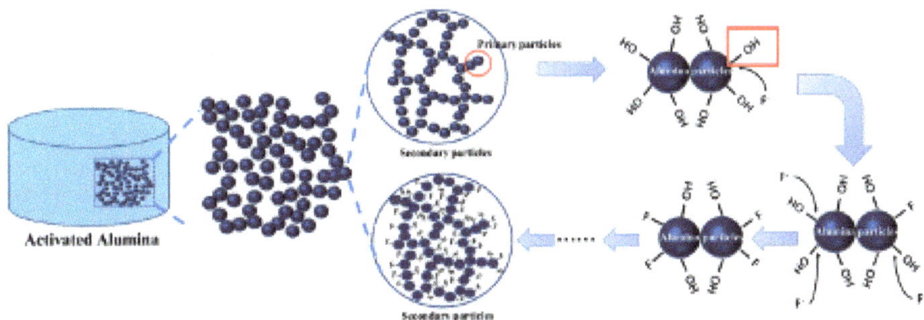

(d) **Flora and Fauna**. Relatively small amounts of fluorides do escape from the process, typically when changing the anodes. And depending on conditions, fluorides could have an impact upon potentially sensitive flora and fauna. However, substantial fluoride reductions have been achieved using new technology and have been demonstrated to pose no threat to health or the environment. A further investment of U.S.\$1.0 Million for a fifty five-hectare buffer zone, in which there would be planted 100,000 trees/plants, had been included in the Alutrint Project. This buffer zone would have provided a physical separation between the community housing and the plant and would have allowed diffusion/dilution of the effluents away.

(e) **Spent Pot Liners (SPL)**. The other waste product generated by Aluminium smelters is the spent pot liners, which absorb fluorides and cyanide during the smelting process. The concern is that, if wet and placed on soil, fluoride and cyanide compounds could have leached into waterways. Alutrint had proposed to mitigate this risk by keeping the SPL dry and either storing it in hermetically sealed containers or exporting it to the U.S. to a licensed disposal facility. Ultimately, Alutrint had attained the environmental standards set by the EMA, including the HF standard for the initial proposal for one pot line. The Certificate of Environmental Clearance (CEC1033/2005) was received on 2nd April 2007.

Evaluating Mr. Dookeran's Comment re Alutrint's "Viability and the optimal use of gas."

(a) **Optimal Use of Gas.** The strategic nature and benefits of the Aluminium industry were very clearly identified by Mr. Panday when the Government approved the Norsk project.

(b) **Viability**. The energy costs were market-related. There was a fixed delivered energy cost when the LME price for Aluminium was less than U.S.\$2100 per tonne and a higher price when the LME was between U.S.\$ 2100 and U.S.\$3000 per tonne. The energy price increased for every incremental U.S.\$100 per tonne in the LME price. The cost of conversion was escalated by a percentage of USCPI from December 2007 for power generated. With combined cycle operation and a special conversion cost for the smelter, the derived gas price, with the pricing structure, was similar to that proposed (and accepted by the Government) with the Norsk project. The annual opportunity cost of the natural gas foregone to the Project would have been U.S.$7.4 million by the allocation of a specific tranche of gas [9] to generate electricity for the smelter.

(c) **Gas utilisation**. In Trinidad and Tobago, the total natural gas consumption was then 4.1 Bscfd, and the Alutrint smelter would have consumed 46.5 MMscfd, which represented 1.1% of total daily natural gas consumption, including the natural gas consumed by the power plant for Alutrint's power requirements. This use of gas in respect of a project to diversify the economy with potential value-added benefits should be compared with the gas utilisation for LNG. Train 1, the smallest LNG plant without any downstream potential, consumed *circa* 480 MMscfd – more than ten times the amount contemplated for the Alutrint project.

The capital requirement (2008) for the project was U.S.$566 Million (excluding U.S.$24 million for a cast house option and U.S.$39 million for contingency). U.S.$400 Million was to be financed through concessionary loan arrangements from the EXIM Bank of China. The remaining U.S.$166 Million represented shareholding equity investments. When compared with other gas-based industries, this capital requirement presented a favourable investment (in terms of CAPEX) per unit quantity of gas utilised.

The average net annual cash flow was estimated at U.S.$45 million with corporation tax of U.S.$11 million per year, and the NPV of the net cash flow, using a discount factor of 8% over a 30-year project life, was estimated at U.S.$425 million. The project ROE would have increased with the incorporation of downstream industries.

(d)   **Downstream Opportunities.** Trinidad and Tobago imported an average of 1000 tonnes of unwrought Aluminium (primarily ingots) annually over the period 2005 to 1Q 2010, which represented about 1% of the output of the planned smelter. Primary products

---

[9] Gas that could be utilised by the Government in lieu of royalties from Amoco.

for secondary Aluminium products would have included Aluminium bars, billets, and ingots, which could have been utilised in moulding, extrusion, and rolling processing units. The consequential economic activity would have been enormous.

Ultimately, the downstream market would have involved value-added products such as automotive and technology parts, packaging, construction applications, and alloys, which take advantage of the properties of Aluminium: it has a light weight, has high electrical and heat conductivity, and it is highly resistant to corrosion.

The Aftermath. Many signed contracts were affected by the cancellation of the Project, including a U.S.$540 million contract with China National Machinery and Equipment Import and Export Corporation. In 2014, Sural, the minority shareholder in the Project, took the Government to arbitration in the United States and made claims (reportedly "upwards of U.S.$100 million"[10]) against the Government for its arbitrary cancellation of the contract.

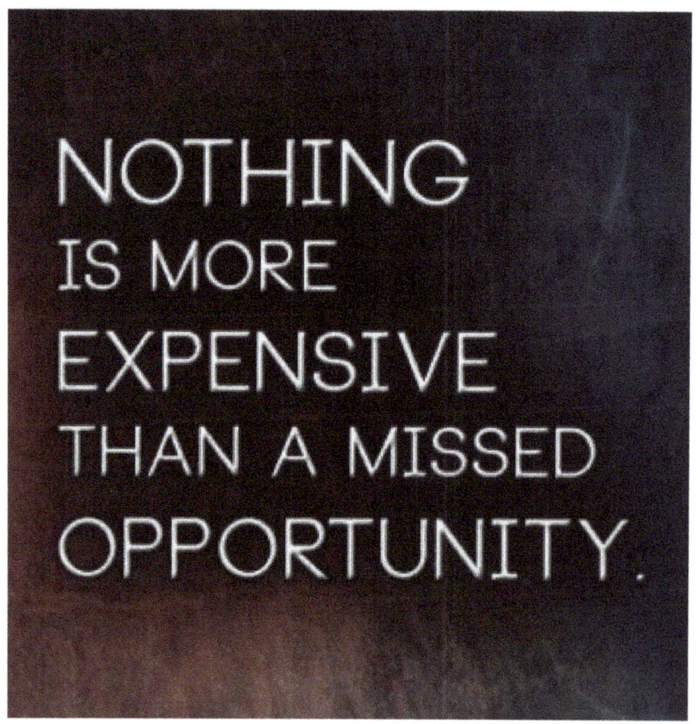

---

[10] According to Prime Minister, Dr Keith Rowley: 'Alutrint Project', *Spotlight*, http://spotlight.lunabyte.io/tag/alutrint-project/Alutrint.

It is reported that later, the Prime Minister, Dr Keith Rowley, told the Parliament's Standing Finance Committee that the Government had entered into discussions with the Chinese government to settle claims for "a much larger sum."

On Wednesday, 3rd October 2018, at the relaunch of the Alutech project,[11] Prime Minister Dr Keith Rowley confirmed that: "We are not restarting the Aluminium smelter. We are not. We have lost it."[12]

---

[11] The aim of this project was to produce Aluminium coils, sheets, and wheel rims and the source of Aluminium would consist of ingots imported from abroad. The project had started several years ago but had been abandoned when the previous administration assumed office. The project is a 60:40 Joint Venture between the Government and Sural.

[12] 'The return of the Alutech plant', *The Guardian*, Wednesday 3rd October 2018, https://www.guardian.co.tt/news/the-return-of-the-alutech-plant-6.2.682666.ee35ec14bd.

Chapter 7: The Challenges of an Ethylene Project

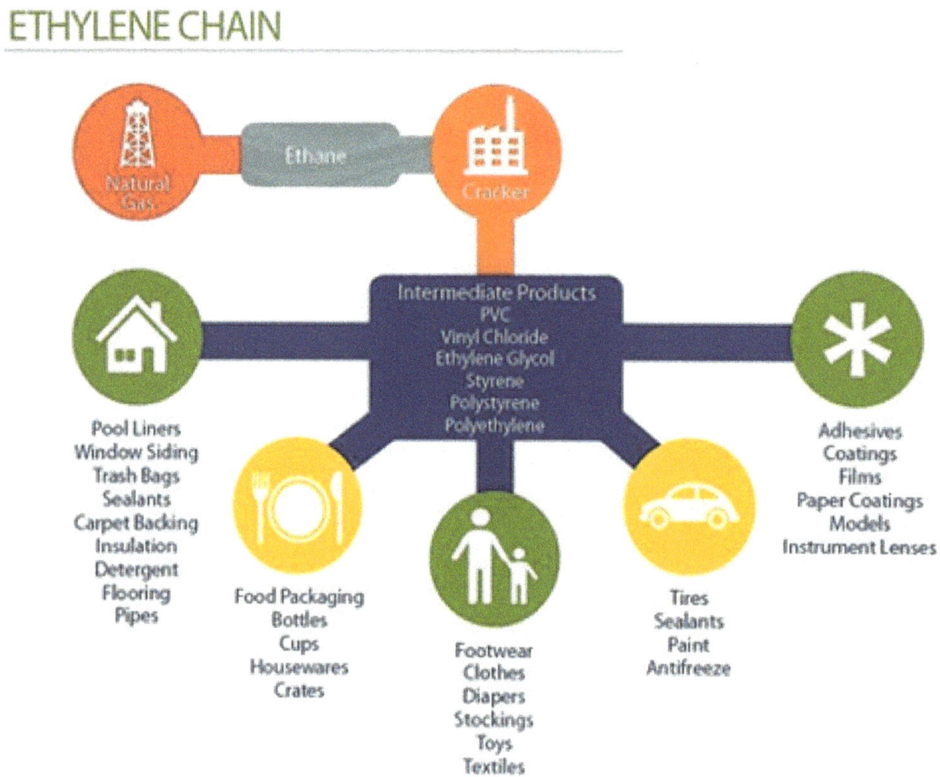

Ethylene is one of the most important petrochemicals in the chemical industry because it is the feedstock for many downstream intermediates, which are the raw materials for many end-use consumer products. Historically, refinery streams have been used to produce ethylene by thermal cracking of the naphtha stream to break large hydrocarbons into smaller ones. In refineries, for example, heptane (C_7H_{16}) can be cracked to produce ethylene (C_2H_4) and pentane (C_5H_{12}).

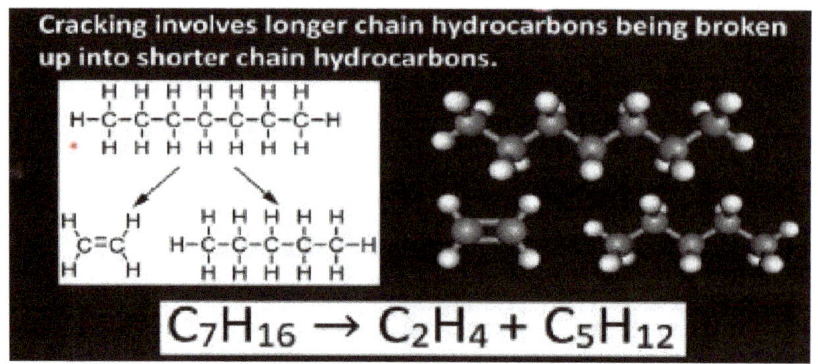

Today, natural gas containing ethane (C_2H_6) is also cracked to make ethylene.

Downstream Intermediates of Ethylene: Polyethylene. Traditionally, the largest outlet for ethylene demand globally has been polyethylene. Low-density polyethylene (LDPE) and linear low-density polyethylene (LLDPE) mainly go into film applications such as food and non-food packaging, shrink and stretch film, and non-packaging uses. High-density polyethylene (HDPE) is used primarily in blow moulding and injection moulding applications such as containers, drums, household goods, caps, and pallets. HDPE can also be extruded into pipes for water, gas, and irrigation, and film for refuse sacks, carrier bags, and industrial lining.

Polyethylene

Applications:

- It is used as a film for packaging food, clothing, and hardware.
- Most commercial trash bags, sandwich bags, and plastic wrapping are made from polyethylene films.
- Polyethylene is also used for everything from seat covers to milk bottles, pails, pans, and dishes.

Ethylene Oxide. The next largest consumer of ethylene is ethylene oxide (EO), which is primarily used to make mono-ethylene glycol (MEG), which is commonly known as antifreeze. Most MEG is used to make polyester fibres for textile applications, PET resins for bottles, and polyester film.

Ethylene oxide is used to manufacture:
- Textiles
- Detergents
- Polyurethane foam
- Antifreeze
- Solvents
- Medicines
- Adhesives
- Sterilant for food and cosmetics
- Sterilization of surgical equipment and plastic devices otherwise damaged by steam

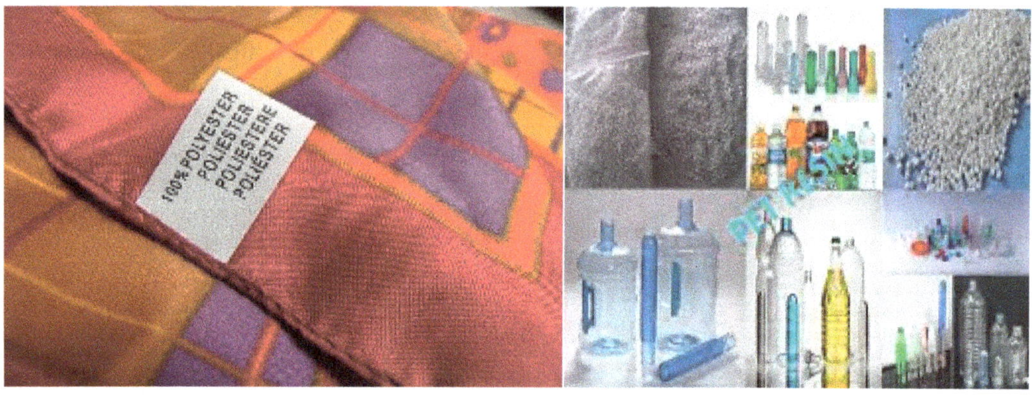

Other well-known end uses of ethylene include –

(a) Polyvinylchloride (PVC).

(b) Styrene.

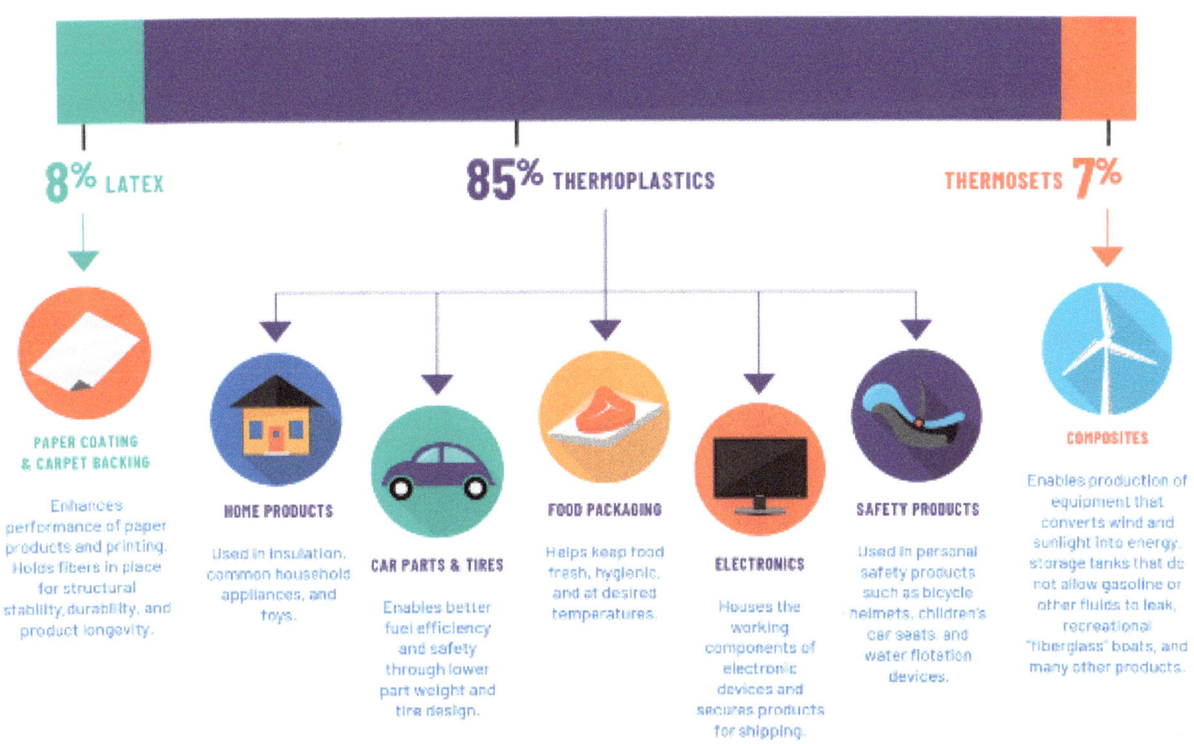

Choice of Feedstock. Ethane, from natural gas, has been increasingly used as an alternative feedstock for the petrochemical industry, and many crackers are based on ethane and natural gas liquids (NGLs). NGLs contain mainly ethane, propane and butanes, and pentane.

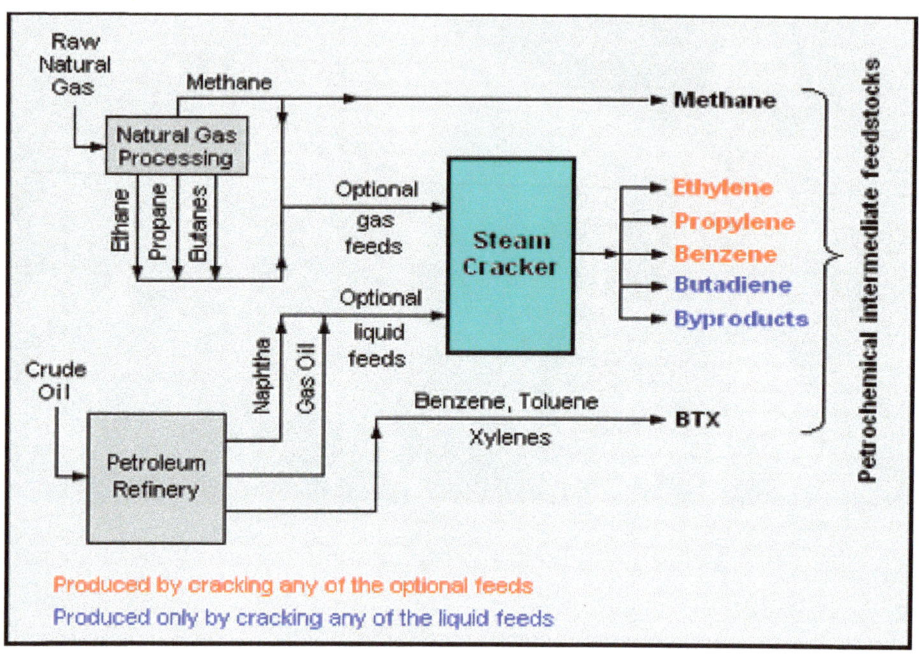

In 1997, NGC and PPGPL studied six feedstock scenarios to determine the most suitable option for a local ethylene cracker.

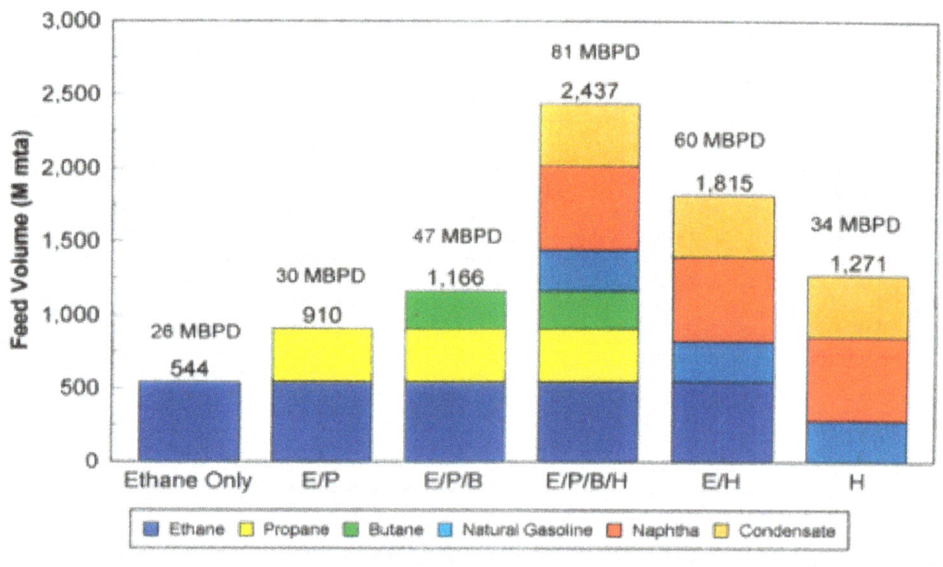

Six feedstock cases were considered

The ethane-only case was the optimum option due to the increases in capital costs of the project and project complexity when NGLs with C_3, C_4, and C_5 components were used as the feedstock.

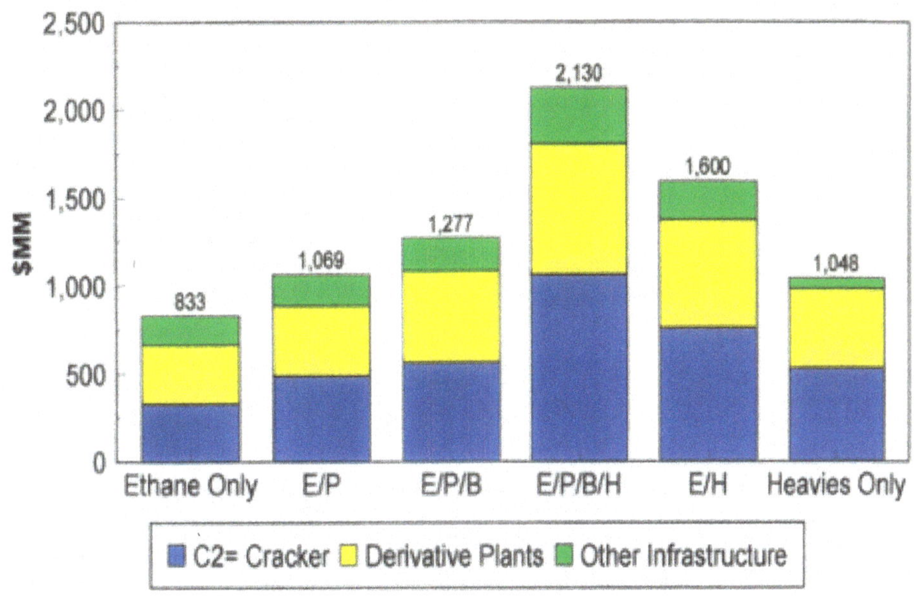

The ethane case is lowest capital option and should be most easily financed

Derivative Plants. Closely associated with ethylene production are derivative plants. These plants are constructed alongside a cracker and convert ethylene from the cracker to a downstream derivative (which is then used to make consumer products). In 1997, Chem Systems recommended ethylene glycol (EG) and polyethylenes (LLDPE and HDPE) as potential derivative plants.

Feedstock Volumes. Two scenarios were considered for the ethane feedstock volumes to supply the steam cracker: Scenario 1, 25,000 Bbls/d and Scenario 2, 40,000 Bbls/d. Under these scenarios, the production of ethylene from the steam cracker would have been either 860 million pounds per year (390 MTPA) or 1,370 million pounds per year (620 MTPA).

Using the ethylene produced from the cracker as feedstock for the derivative plant, two options were considered. Under Case A, the derivative plant would be designed to produce polyethylene and ethylene glycol, and under Case B, polyethylene only. The quantities of ethylene and the attendant derivatives based upon these Cases and Scenarios are illustrated in the table below.

Ethane Volume (Bbls/d)	Ethylene Production MPA (MTA)	Derivatives	Derivative Production MPA (MTA)
Case A: Polyethylene and Ethylene Glycol			
25,000 (Scenario A)	860 (390) *	LLDPE/HDPE	460 (210)
		Ethylene Glycol	630 (285)
40,000 (Scenario B)	1370 (520) *	LLDPE/HDPE	735 (335)
		Ethylene Glycol	1000 (455)
Case B: Polyethylene Only			
25,000 (Scenario A)	860 (390)	LLDPE	920 (415)
40,000 (Scenario B)	1370 (520)	LLDPE	1,470 (665)

* 50:50 split of the ethylene production is used for the production of derivatives.

The target markets and netback pricing were identified as the Far East for ethylene glycol and the Far East, Latin America, USGC, and Western Europe for polyethylene.

Ethane Price to the Ethylene Complex. The calculated highest ethane price which the complex would have been able to pay (assuming it needed an IRR of 17%) depended upon both the available ethane volume and the derivative configuration of the project. This variation is illustrated in the table below for the previously identified scenarios for ethane volumes and derivatives.

Ethane price to achieve a 17% IRR for the Project under different Scenarios

Ethane Volume (Bbls/day)	Derivatives	MTPA	Ethane Price (¢pg)
	Polyethylene	460	9.4
25,000	Ethylene Glycol	630	

	Polyethylene	735	
40,000	Ethylene Glycol	1,000	18.6
25,000	Polyethylene	920	19.6
40,000	Polyethylene	1,470	27.6

The U.S. Gulf Coast ethane forecast market price (the year 2003) was 26.9 cents per gallon.

Conclusions. The studies led to the following conclusions:

(a) The Trinidad complex would have to compete with countries with an "advantaged ethane price," namely, the Middle East, Western Canada, and Venezuela.

(b) A complex producing only polyethylene was likely to be significantly more profitable than one producing polyethylene and ethylene glycol.

(c) The profitability of the project would increase by selling products preferentially in Latin America rather than the more competitive markets like Southeast Asia.

(d) Larger plants would be more profitable.

Accordingly, a sufficient volume of ethane and an appropriate ethane price were essential for a local project to be successful.

Ethane Volumes. The ethane feedstock rate to most ethane-based ethylene plants then in operation around the world was between about 25,000 Bbls/d and 60,000 Bbls/d. Proposed plants, or those under construction, had ethane feedstock rates of about 40,000 Bbls/d to 60,000 Bbls/d. When three natural gas production scenarios were considered in Trinidad and Tobago, the potentially recoverable ethane volumes were projected to be between 23,000 Bbls/d and 46,000 Bbls/d. And, a project could be economic with an ethane feedstock rate of 40,000 Bbls/d but marginally so with 25,000 Bbls/d.

Uncertainties. The main uncertainties when assessing the ethane volume forecast were:

(a) The gas demand;

(b) The ethane content of the gas produced; and

(c) The quality of the gas added to gas reserves.

Gas Demand. In 1997, total gas demand was 700 billion BTU per day and was expected to double by 1999. Looking beyond 1999, the gas demand was projected under (a) most likely; (b) optimistic; and (c) pessimistic scenarios. The optimistic scenario assumed two LNG Trains.

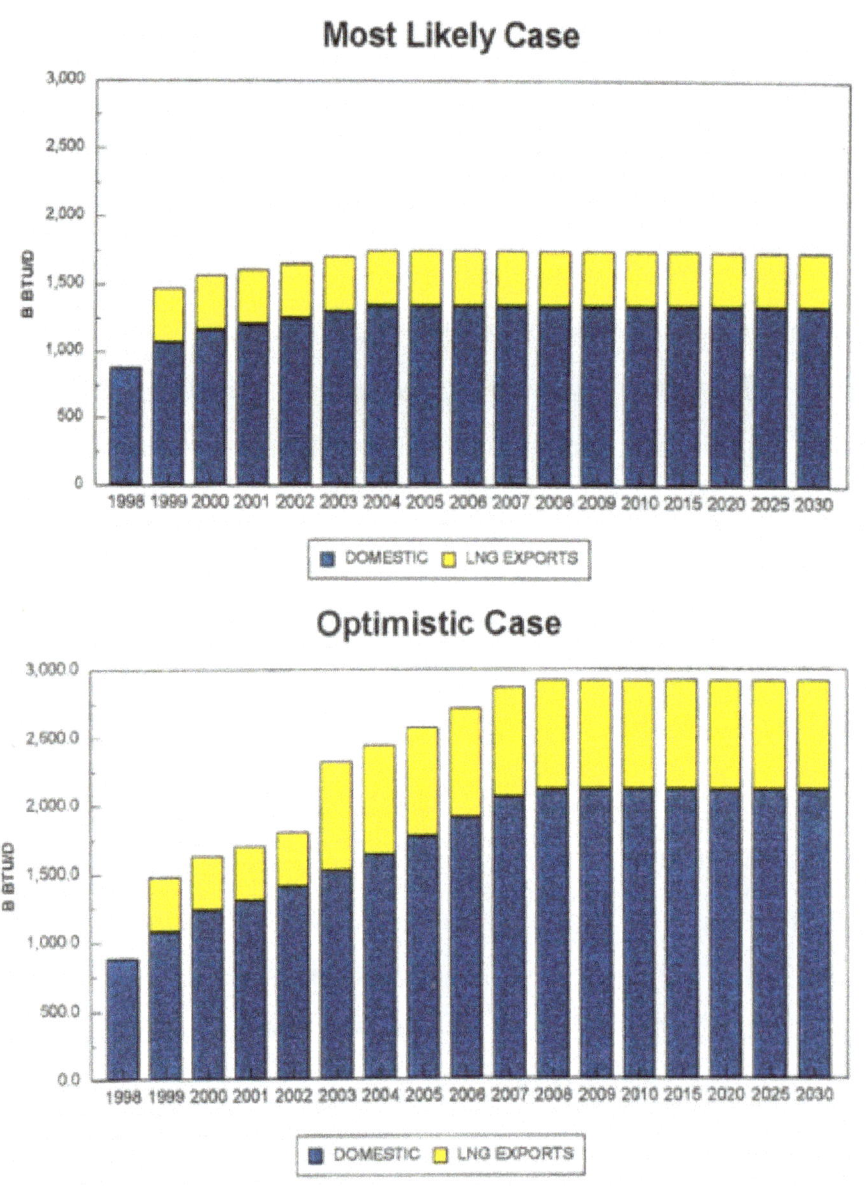

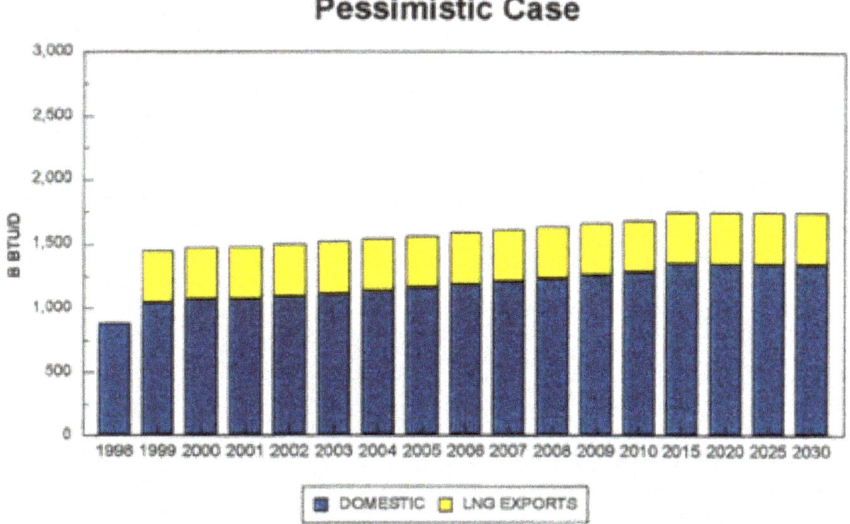

Gas quality. The gas composition has an impact upon ethane availability and the cost to extract ethane. It was estimated that 21 Tcf of ethane-rich reserves would have been needed in order to justify a 20-year, 40,000 Bbls/d project. And it was believed that, out of Trinidad and Tobago's 22 Tcf of then-current gas reserves, only 14 Tcf had contained economically recoverable ethane. The commingling of gas from multiple sources, the depletion of reservoirs, and the connection to new fields had also made it difficult to predict how the aggregate gas composition would have varied over time. The map below shows the gas that is produced from the different fields in Trinidad and Tobago off the North Coast ("NCMA") and the East Coast ("ECMA"). Each field had its own unique gas quality.

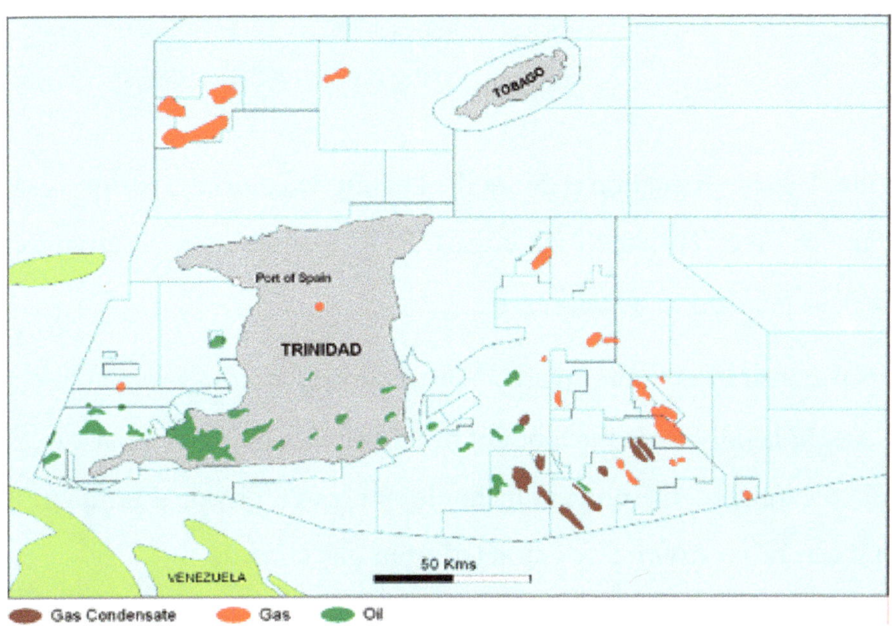

Atlantic LNG. The ethane content of gas supplied to Atlantic LNG was forecast to become leaner because of the commingling of the gas from the ECMA with the dry gas from the NCMA. Atlantic LNG Train 1 was to be supplied with ethane-rich reserves from the Mayaro East and *S/SE* Galeota fields. However, a second LNG train, which was assumed to start up in 2003 in the optimistic case, was to be supplied with 50% of the ethane-rich gas reserves from the ECMA and 50% of the dry gas from the NCMA. By comingling the gas from the NCMA and the ECMA for Train 2, the resulting ethane content of the feedstock would have decreased.

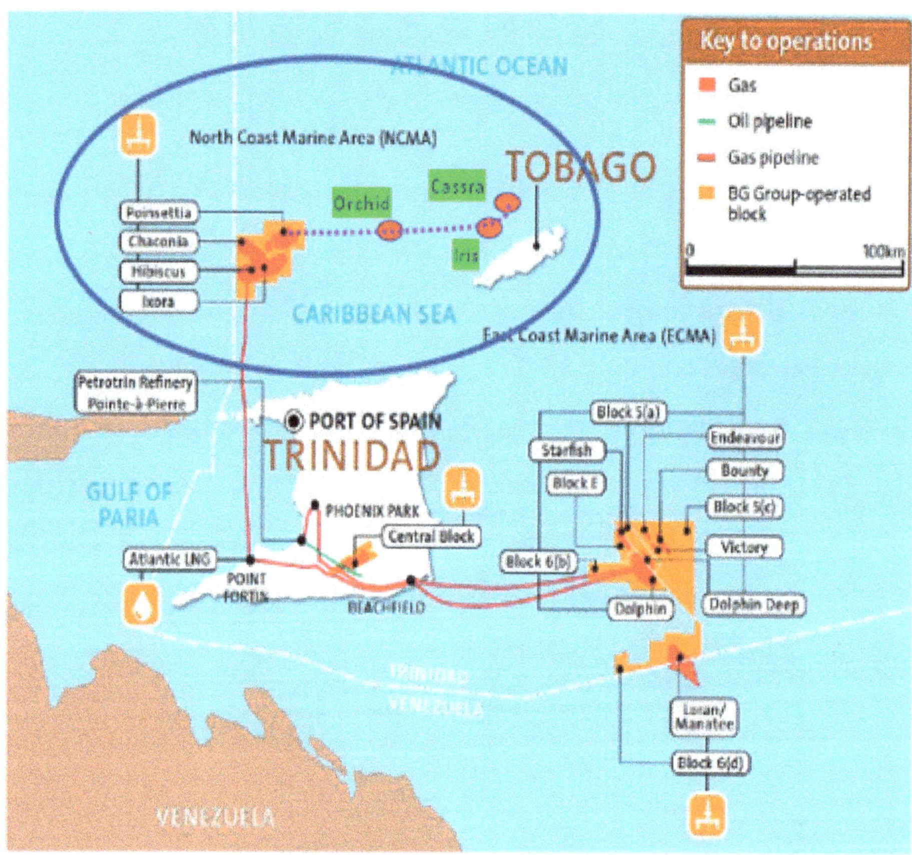

Ethane Extraction. Ethane extraction reduces the heating value of the residue gas. This reduction in the BTU content, or Plant Thermal Reduction ("PTR"), sometimes referred to as the "shrink," has two consequences that had to be addressed:

   (a) Commercial constraints – the quality specifications of the product for customers, in practice, would legally limit the amount of ethane that may have been extracted; and

   (b) Technical constraints – the process technology created physical limits to the amount of ethane that can be economically extracted from gas streams.

NGC's Contracts. Many of NGC's gas sales contracts had a minimum heating value specification of 1,000 BTU per cubic ft., but NGC's new sales contracts had introduced a minimum heating value specification of 970 BTU per cubic ft. At PPGPL, with 90% ethane extraction, the heating value of the residual gas stream would have been reduced from 1,033 BTU per cubic ft. to *circa* 1,005 BTU per cubic ft., which could have been tolerated within NGC's specification limits.

ALNG Contracts. Under the Train 1 LNG offtake contract with Enagas, the LNG, when delivered by Atlantic 1 was required to have *"a Gross Heating Value of not less than 1030 BTU per Standard Cubic Foot and not more than 1085 BTU per Standard Cubic Foot."* The minimum and maximum values in the Cabot contract were specified to be 1000 and 1075 BTU per scf, respectively. Ultimately, if it were acceptable for the LNG customers to accept LNG with a heating value as low as 1,015 BTU per standard cubic ft., then it was believed that, in a worst worst-case scenario, Atlantic 1 could economically extract at least 65% of the ethane from the gas that had been delivered at Point Fortin.

Ethane Pricing. During the project development, the team adopted a concept of an ethane "ceiling price," which was a market-based price that represented an estimate of the *"highest price a buyer could reasonably pay"* for ethane. In that regard, ceiling price estimates were made using the most economically attractive scenarios, i.e., ethane-based feedstock volumes to make ethylene and the polyethylene derivative. These scenarios resulted in the following ethane ceiling prices:

Ethane Feedstock Rate	25,000 Bbls/d ethane	40,000 Bbls/d ethane
Ethylene Capacity*	390,000 MTA	620,000 MTA
Ethane Ceiling Price** (1997$US)	$0.169/gallon	$0.238 /gallon

* At 80% conversion & 90% utilisation.   ** Ten-year tax holiday assumed.

The difference in ceiling prices between the two scenarios for ethane feedstock (almost 7 ¢pg) showed that an investor in a 40,000 Bbls/d complex could have afforded to pay 40% more for ethane feedstock than with a 25,000 Bbls/d complex. This demonstrated the economies of scale enjoyed by the larger facility when the ceiling price was being considered. In other words, the availability of larger volumes of ethane would have provided greater flexibility during pricing negotiations with a proposed investor in an ethylene complex.

Ethane Floor Price. Another concept, the ethane "floor price," was a cost-based price that represented the lowest price at which ethane could have been recovered or produced by the seller of ethane. The floor prices for three study cases are shown in the following table:

	Case A	Case B	Case C
Average Ethane Rate Over 20 Years	25,000 BPD	29,000 BPD	43,000 BPD
Ethane Floor Price (uniform annual cost basis,1997$US) *	$0.202/gallon	$0.191/gallon	$0.172/gallon

* No tax holiday assumed.

The difference in the floor prices between Case A and Case C was $0.019/gallon ($\approx$ 3 ¢pg), which demonstrated lesser economies of scale than that observed with the scenarios for the ceiling price.

Comparison of ceiling and floor prices. Another metric that was used to determine the relative attractiveness of an ethane project was an evaluation of ceiling prices in relation to floor prices. For the project to be economically attractive, the ceiling price for a potential investor would have to be greater than the floor price at which ethane could have been produced by PPGPL.

For example, if the lowest price that PPGPL could offer an investor for ethane was 15 ¢pg, the project would not be attractive if the highest price the investor could pay was 10 ¢pg. The comparisons of ceiling prices and floor prices are shown graphically for the most likely and optimistic ethane volume cases, along with the impact of the sensitivity analyses.

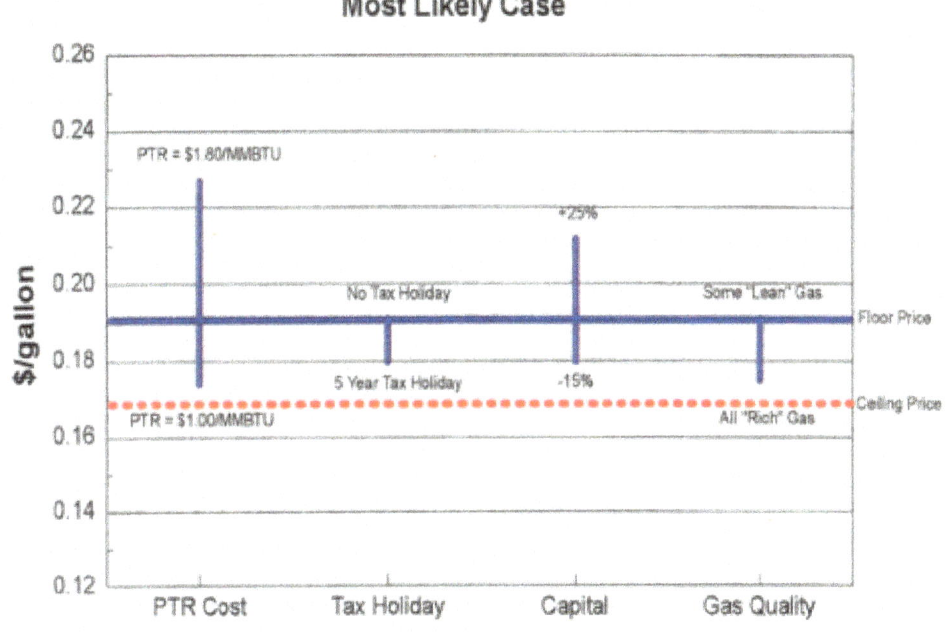

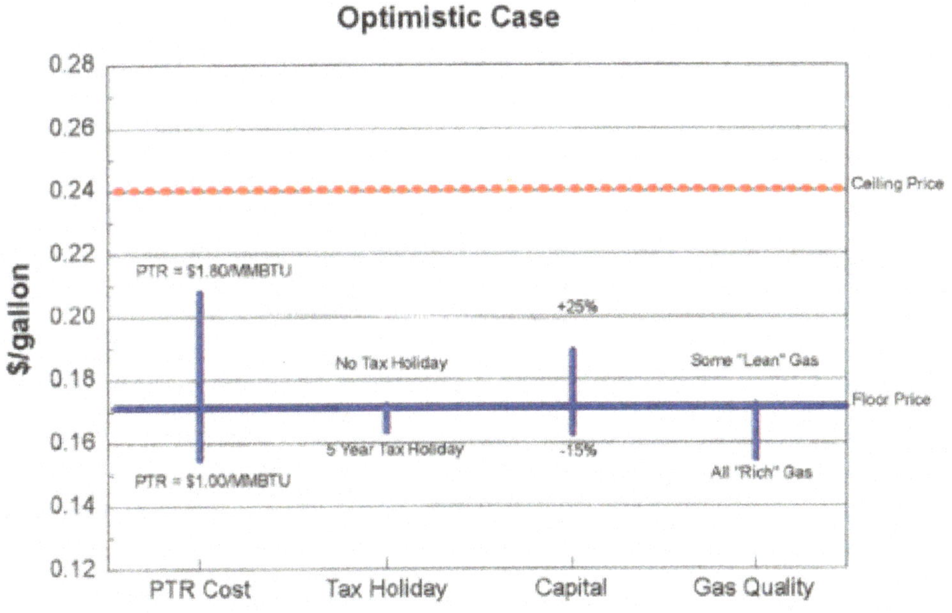

Under the most likely case scenario – one LNG Train - the floor price was greater than the ceiling price. Thus, it would have been necessary to combine a significant reduction in the PTR with favourable tax incentives in order to improve the economic attractiveness of this scenario. In contrast, under the optimistic case with a total of two LNG Trains, the ceiling price was 24 ¢pg compared with a floor price in the vicinity of 17 ¢pg. The difference (about 7 ¢pg) indicated that at higher volumes of ethane availability, the project would become economically attractive.

Proposed Investors. Following a request for proposals, NGC/PPGPL had received two commercial proposals from two major polyethylene producers. Investor A submitted a proposal that had expressed a willingness to restrict operations to 70% capacity in year 1 (22,000 barrels of ethane), increasing to 90% and full capacity in years 2 and 3, respectively. However, the proposal had lacked firm commitments regarding an investment decision. This investor later withdrew. Investor B's proposal had included the following elements –

(a) A base case feedstock supply of 35,000 barrels of ethane in 2003;

(b) A 550,000 tons per annum ("500 KTA") ethylene cracker producing 250 KTA of HDPE and 300 KTA of LDPE;[13]

(c) Minimum ethane requirements of 24,000 bpd that could be supplemented with 20% propane for a total volume of 36,000 barrels if ethane were to be in short supply.

(d) A feedstock pricing mechanism that was margin-sharing based upon a weighted average netback of primary products.

(e) A proposed ethane price of 9.3 cents per gallon, based upon a product market price with floor and cap concepts.

Since PPGPL had determined the ethane processing fee to be 9.2 cents per gallon or $1.40/MMBTU, the proposal from Investor B seemed to be promising.

Atlantic LNG. In anticipation of the negotiations with the shareholders of Atlantic 1 and the Government in relation to the proposed Atlantic LNG Train 2/3 expansion, Gordon Shearer wrote the Minister of Energy and nominated me to be Cabot's representative on the Atlantic 1 negotiating team. However, in light of my earlier involvement with PPGPL on the development of an ethylene project, I formally declined Cabot's nomination and later wrote the Minister accordingly.

Prior to the negotiations with the shareholders in Atlantic LNG, the Minister of Energy called a meeting of the various stakeholders in the public sector. At that meeting, Mr. Clinton ("Billy") Rambaransingh, the Chairman of PPGPL, and I were introduced as the representatives for PPGPL

---

[13] According to the proposal, that approach had minimised capital costs while still providing world-scale HDPE and LDPE production.

in respect of all matters concerning ethane and NGLs to be made available from the Atlantic LNG facility. Ultimately, a team, which included representatives from NGC and PPGPL collectively (the "PPGPL Team"), conducted the negotiations with the ALNG team at the Ministry's offices at Riverside Plaza in Port-of-Spain.

The Atlantic LNG Negotiations. The PPGPL Team made proposals to Atlantic LNG in accordance with the following negotiation objectives and strategy:

(a) **Ethane Aggregation**. At the time of the negotiations, PPGPL had been processing all the gas that NGC had aggregated for sale to the domestic market. PPGPL had also processed the NGL stream that it had obtained from Atlantic 1 during the LNG manufacturing process for Train 1. PPGPL had, therefore, been operating as an NGL processing hub in Trinidad and Tobago.

Accordingly, the PPGPL Team proposed that ethane aggregation and processing should occur at a single location in Trinidad, namely, at PPGPL's plant, because it would have been an incremental NGL processing step, which would have reduced ethane processing costs.

(b) **Ownership of Ethane.** Since the ethane in the gas stream that was processed by PPGPL was currently owned by NGC, the PPGPL Team believed that NGC should also own any ethane that was transferred to PPGPL at the delivery point in Point Fortin.

The rationale was to ensure single-point responsibility in discussions with prospective investors.

(c) **Contractual Term**. Considering the ethane feedstock supply requirements for an ethane cracker, which was anticipated to be at least twenty years, the PPGPL Team believed that the term for the ethane supply to PPGPL from Atlantic LNG should coincide with the term for the Train 2/3 LNG Expansion.

(d) **Ethane Pricing**. Given the uncertainty in ethane availability, it was prudent for the PPGPL Team to negotiate pricing terms that would satisfy the economics of a more costly project that involved mixed feed options, for example, ethane and propane (and even butane).

Thus, the PPGPL Team proposed, among other things, that, as an "enabler" for the 2-Train LNG Expansion, Atlantic 1 should make ethane available to Trinidad and Tobago as part of the NGL stream at a *de minimis* price[14]. The rationale was that the issues relating to ethane were not merely commercial but were of National importance. In essence, since ethane was a strategic raw material for Trinidad and Tobago, any costs incurred in removing it in furtherance of the Train 2/3 Expansion, should be borne by the investors in that project.

Ultimately, under the worst-case scenario, the PPGPL Team would have been willing to accept that the price of ethane to be purchased from ALNG was to be based upon its BTU value and **not** the market value of the ethane if it were to be sold as LNG or ethane.

(e) **Ethane Extraction**. It was critical for the PPGPL Team to maximise the amount of available ethane that could be guaranteed to a potential investor. Thus, the PPGPL Team sought a commitment from the ALNG shareholder team that no less than 65% of the ethane in the gas stream supplied to Atlantic LNG should be made available to PPGPL in the NGL stream.

The Game Changer. Out of the blue, on 28th September 1999, the Minister of Energy and Energy Industries wrote the principals of Atlantic 1 in relation to the ethane negotiations and indicated that "the negotiations with regard to ethane and NGLs are to be conducted by the Government's team." Since the Government's team was the team that had been conducting the overall negotiations for the Government in respect of the Train 2/3 Expansion, the PPGPL Team had been removed and thereafter played no further part in the negotiations with Atlantic LNG.

---

[14] The initial proposal was "no cost" but was later modified to U.S.$0.01.

Mr. Gerry Peereboom,   Mr. David Wight,
President, Atlantic LNG   Chairman, BP Amoco

Mr. Jacques Robinson,   Mr. Carlos Quintana,
President, British Gas   Repsol LNG

Dear Sirs,

<u>Negotiations for Two-Train Expansion of
LNG Facility – Ethane Issues</u>

Reference is made to the current negotiations for the expansion of LNG facilities by two additional trains as proposed by the Shareholders of Atlantic LNG.

You will no doubt recall the discussions held on September 6 and 13, 1999 and the importance which the Government of Trinidad and Tobago has placed on the development of an ethane based Ethylene Cracker and related downstream conversion facilities. This facility is viewed as critical to the development of downstream value-added gas based industries which would bring significant sustainable employment opportunities and permit the development of greater economic activities based on the active participation of local industry in the petroleum and petrochemical sector. I have stated that approval of the proposed two-train expansion of the LNG facility is contingent on the availability of ethane and natural gas liquids at a price which would permit the development of an economically viable ethylene complex.

It is important to recognize that the pricing of the Ethane and NGL's is critical to the success of the country's efforts to attract investments in the ethylene facilities. I need to emphasize the importance of employment creation industries as quid pro quo for the depletion of a non-renewable natural resource as represented by a two-train LNG project. In this regard, the Government has set negotiations with the potential investors for an ethylene complex for Trinidad and Tobago for the second half of October, 1999 and as a result, an attractive response for final and definitive proposals for Ethane and Natural Gas Liquids pricing and availability by October 14, 1999.

Further, I wish to reiterate that all negotiations with regard to Ethane and NGL's are to be conducted with the Government's team.

Yours faithfully,

(Finbar K. Gangar)
Minister of Energy and Energy Industries

In March 2000, the Government approved a Heads of Agreement for the expansion of the facility at Point Fortin, which set out various terms for the LNG Expansion, including an "Ethane Term Sheet." The terms for ethane in the Ethane Term Sheet were reproduced in the Atlantic LNG 2/3 Expansion Agreement (the "Expansion Agreement") dated 7th December 2000 between the Government and Atlantic LNG 2/3 Company of Trinidad and Tobago (the "Company").

The Expansion Agreement. Article 6 of the Expansion Agreement contained the following provisions relating to ethane:

(a) Section 6.1 (Sale of Ethane).

"Subject to the conditions ... in Section 6.3, the Company shall sell the .. Ethane Stream separately or as part of a mixed stream with other NGLs to a Government nominated entity (the "Ethane Buyer") ... pursuant to an agreement between the Company and the Ethane Buyer ... which will include, among others, the terms described in an Ethane Term Sheet attached hereto as Exhibit 6.1."

(b) Section 6.3 (Conditions Precedent).

".... (b) By December 31, 2005, the Ethylene Project Investor must have provided the Company with evidence of an affirmative final investment decision to construct, own and operate an ethylene plant in Trinidad, utilizing the Ethane Stream (the date of such final decision to be the "Ethylene Investment Decision Date"). Within thirty (30) days after the Ethylene Investment Decision Date the Ethylene Project Investor shall give to the Company written notice specifying the start up date of the ethylene plant, such start up date to be not less than thirty (30) months after the date of such notice. After receipt of such notice the Company shall proceed with the installation of suitable facilities for the extraction of ethane in the Additional Trains in order to deliver to ethane in a timely basis."

(c) Section 6.6 (Failure to Satisfy Conditions):

"In the event that the Ethylene Project Investor does not take a final investment decision as contemplated by section 6.3(b) by December 31, 2005, the provisions of Section 6.1 to 6.5 hereof inclusive shall cease to be binding on the Parties ... ."

Commercial Analysis. The terms of the Expansion Agreement relating to the supply of ethane can be distilled down to two key points:

(a) The Atlantic LNG 2/3 Company's obligation to supply ethane to an Ethylene Project Investor was only within an approximate 5-year window that closed on 31st December 2005.

(b) If the Ethylene Project Investor did not make a final investment within that 5-year window, the terms agreed in the Ethane Term Sheet were no longer binding on the Parties.

The figure below demonstrates that the volume of ethane available for an ethylene project was the most critical variable that had impacted the NPV in the project economics for an ethylene plant.

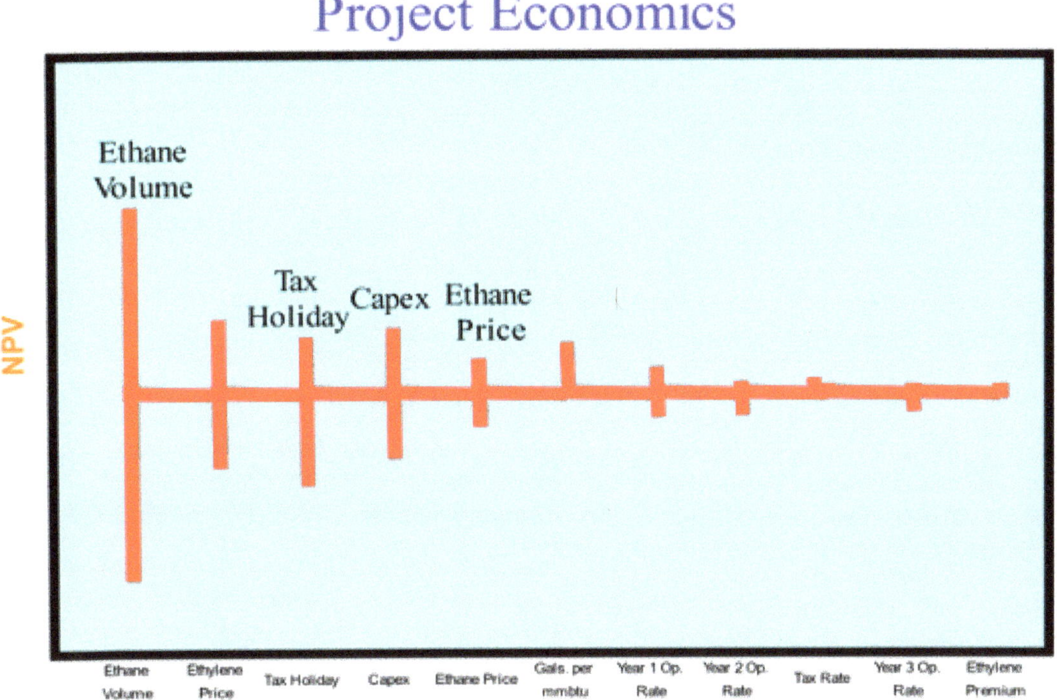

Given the importance of ethane volumes and the uncertainty in the timing of an investment in an ethylene project, why was the LNG 2/3 Company let off the hook after 31st December 2005? Indeed, in light of the Minister's acknowledged importance of an ethylene project and that approval of the Expansion was *"contingent upon the availability of ethane,"* it is difficult to comprehend the rationale for his acceptance of a time limitation for making ethane available from the LNG Facility.

The Ethane Term Sheet. The relevant provisions of the Ethane Term Sheet are set out below.

(a) <u>Section 2.1.</u>

"Subject to Section 4, Seller shall reasonably endeavour to install equipment at Seller's Plant to extract Ethane and commence deliveries of an Ethane Stream to Buyer on the contemplated start-up date."

(b) <u>Section 4.1.</u>

" ...While Seller believes [the] modelling work to be the best valuable predictor of the LNG Facility's performance, it will not be construed as any sort of guarantee of actual ethane volumes, and it is recognized that the actual results will vary as gas composition changes over time."

(c) <u>Section 4. 2.</u>

"Seller shall have the sole discretion to control the operations of Train 2 and Train 3 and to determine what percentage of the total quantity of the Ethane Stream shall be recovered during such operations. Notwithstanding, Seller shall use reasonable endeavours to optimise the recovery of the Ethane from the natural gas delivered to Train 2 and Train 3 within the physical constraints of such Trains while still maximising the overall output on an energy basis and meeting its contractual obligations to the buyers of LNG and NGLs from Train 2 and Train 3."

Thus, the Atlantic LNG 2/3 Company had provided very guarded legal obligations to make ethane available to an Ethylene Project Investor.

Accordingly, given the strategic nature of an ethylene project, why did the Minister agree that LNG production should be given priority over the supply of ethane from LNG processing when a critical objective in his letter to shareholders was to ensure that Train 2/3 project approval was subject to ethane availability?

Relevant Perspectives. The security of gas supply feedstock commitments that Amoco had to provide to the shareholders in Train 1 in respect of the Train 1 LNG Plant provide a proper perspective of the commercial terms that the Minister had agreed in respect of ethane supply in the 2/3 Expansion Agreement.

For the Train 1 LNG project, Amoco had given commitments that –

(a) The dedicated gas fields would contain sufficient reserves to enable Amoco to perform its gas supply obligations under the 20-year contract;

(b) 4.8% of the proved reserves would remain at the end of the 20-year contract;

(c) Amoco would evaluate probable and possible reserves to provide upside to the then-existing levels of proved reserves dedicated to the contract;

(d) Amoco would maintain deliverability in excess of its contractual requirements as a cushion against any unforeseen operational problems; and

(e) Amoco was liable to Atlantic 1 in respect of any loss or failure of the relevant gas reservoirs and the deliverability associated therewith due to natural depletion or the absence of economically recoverable gas (other than such an absence which itself was the result of *Force Majeure,*).

By contrast, the terms of the Expansion Agreement, with respect to ethane, expressed the following objectives:

(a) **Priority.** LNG production by the Atlantic LNG 2/3 Company had priority over ethane recovery;

(b) **Ethane Volumes/Guarantees**. The estimated ethane obtained from the plant was not to be construed "as any sort of guarantee of the ethane volumes" to be supplied by the Atlantic LNG 2/3 Company; and

(c) **Obligation Period**. There were no legal commitments from the Atlantic LNG 2/3 Company to supply ethane to any investor if an investment in an ethylene project had not been made after 31st December 2005.

The foregoing considerations exacerbated the concerns that had already existed about the volumes of ethane that might be available from indigenous gas fields to make an ethane-based ethylene project feasible.

Thus, it is highly questionable whether, after the appropriate due diligence, the Board of an Ethylene Project Investor (and/or its financiers) would have made an investment decision to proceed with an ethylene project costing circa 1 billion U.S.$.

Accordingly, any attempts to implement an ethylene project in Trinidad and Tobago were likely to be frustrated after the Expansion Agreement was signed in 2000.

Technical Considerations. As regards the separation of ethane from methane, due to the critical temperature of ethane (90.1 °F), conceptually, ethane can normally be separated from methane (critical temperature, - 117 °F) during cryogenic gas processing to produce an NGL stream containing a mixture of ethane, propane, etc. This would have occurred at the PPGPL facility.

Notwithstanding differences between cryogenic processing to produce LNG or NGLs, the sub-optimal decision-making with respect to ethane availability of ethane from the Train 2/3 Expansion is borne out by the terms of the subsequent LNG Train 4 project agreement with the Government. In the later agreement, the parties had agreed that "the design of Train 4 would include the technical capability for the installation of the necessary equipment to extract ethane up to a maximum of 80% of the inlet ethane content based on design feed gas composition with actual extraction levels dependent on actual feed gas composition and ethane extraction equipment scope."

Yet another big fish got away!

Chapter 8: The Role of NGC

Overview. During my tenure as Vice President, Business Development, the company's corporate Mission was to maximise value from the development of the natural gas industry for the benefit of Trinidad and Tobago. Ultimately, therefore, NGC had aimed to secure dual roles of profitable operation for the company as well as maximum economic benefit to Trinidad and Tobago. Until the implementation of Train 1 at Atlantic LNG, NGC was the sole purchaser of natural gas from offshore producers (an "aggregator") and the sole seller of natural gas to consumers (a "wholesaler"). I led the teams that negotiated gas supply purchases from upstream gas producers and gas sales to the downstream export projects. My Group consisted of three areas:

(a) The project planning division, which identified and selected value-added natural gas-based petrochemicals for local manufacture. This function was previously performed by the NEC before it merged with NGC.

(b) The infrastructure planning division, which was responsible for the development of new industrial estates, new port development, and the management of NGC's land and marine assets such as the tugs and workboats and the Savonetta piers.

(c) The strategic planning department, which increased gas sales to the light industrial and commercial sector, including sales of gas for CNG. The department aimed to make natural gas the premier fuel within the domestic economy, with a motto: "Sell More Gas!"

Today, the Infrastructure Planning and the Project Planning functions are carried out by the NEC (now rebranded as "National Energy"), which is a wholly-owned subsidiary of NGC.

Under a contract with NGC, PPGPL, a midstream cryogenic gas processing facility, in which NGC has held the majority interest, processed all the gas that NGC aggregated from upstream gas producers and the associated gas from NGC's offshore platforms at the Teak and Poui Fields.

NGC, as a wholesaler, sold the processed gas (mainly methane and ethane) for electricity generation, for the light industrial and commercial sectors, and for the export industries, including petrochemical manufacture.

Today, PPGPL also processes NGLs under a contract with Atlantic LNG and exports the resulting NGLs (propane, butane, and pentane, and heavier hydrocarbons).

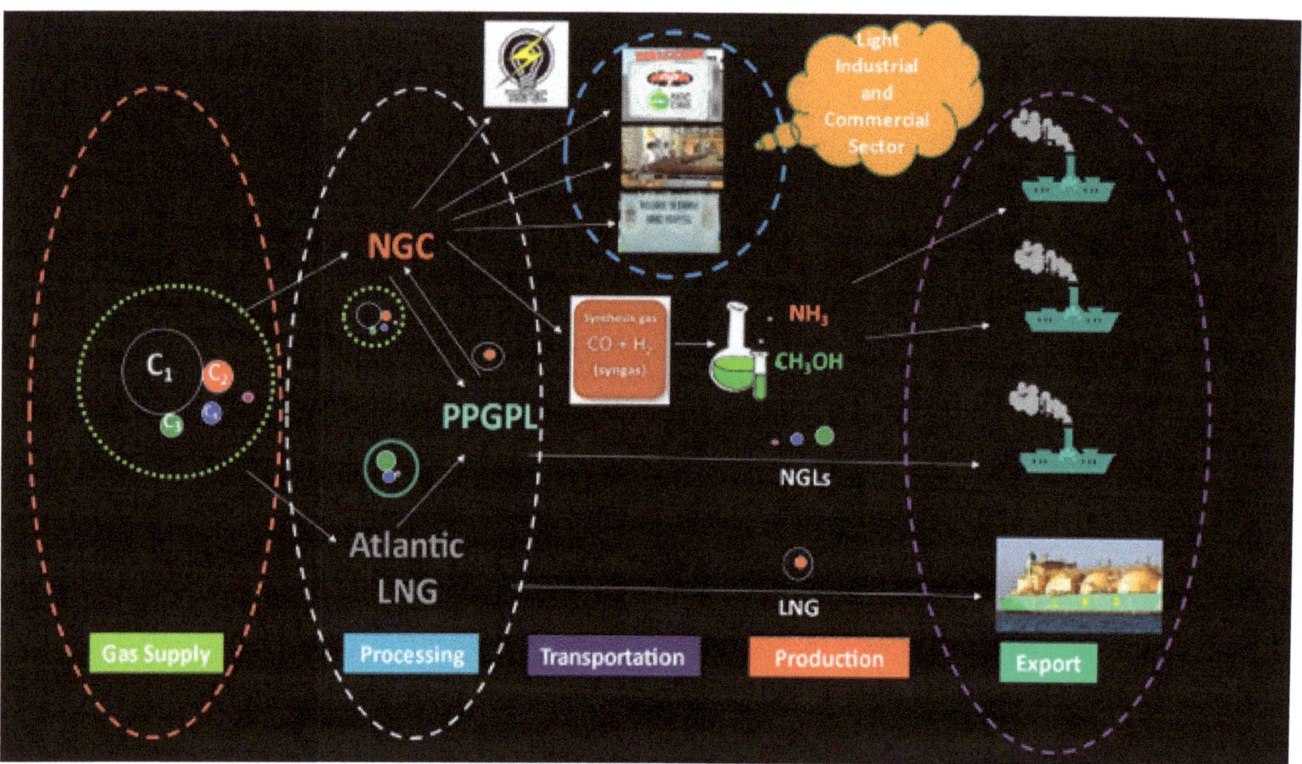

Current NGC Roles. NGC's current roles in the energy value chain include the following:

(a) Ownership interests in local upstream assets through wholly-owned subsidiaries;

(b) Aggregator of gas from upstream sources and the sole wholesaler of gas to downstream export industries at Point Lisas, T&TEC, and the light industrial and commercial sector;

(c) The owner and operator of the onshore transmission and distribution system and the sole transporter of gas to the downstream sector;

(d) Majority interests in PPGPL, which controls the availability of NGLs for possible downstream manufacturing operations; and

(e) Through National Energy, the promotion and development of new projects in the sector, which can involve recommendations for fiscal and other incentives.

Should the status quo be changed? Implicit in the question is whether the current arrangements still best serve the interests of Trinidad and Tobago, considering the changes in the natural gas environment both locally and globally. The issues are not straightforward. For example, should changes be made in the midstream with NGC without making changes in other parts of the value chain? What legislative requirements, if any, would be required? In the light of the liberalisation of the gas sectors in the U.S. and Europe (particularly the U.K.), should the energy sector in Trinidad and Tobago be similarly liberalised? Has NGC's current monopolies led to conflicts of interest that have gone unchecked without independent regulation?

In August 1997, as the Vice President, Business Development, I had to address some of these issues when interviewed by World Bank personnel (the "WB Team") who had visited NGC, during a process of consultation, with the aim of making recommendations to the Government in respect of NGC's role in the energy sector. At the time, I did not believe that the WB Team had sufficiently understood the local environment, which was different in many respects from the prevailing conditions in the developed world. Accordingly, after hearing their views, I explained to them that extreme caution had to be taken when proposing solutions that were relevant to other countries with different policy and/or political considerations. In any event, ultimately, the WB Team produced a "Trinidad and Tobago Gas Sector Strategy Note" dated August 14, 1997 ("the Strategy Note") in which arguments had been made to the Government to change the role of NGC.

The Strategy Note. From time to time, some aspects of NGC's role in the energy sector had been challenged, and the Strategy Note,[15] I believed, had been the result of lobbying by some of the major players in the sector - An upstream producer might see NGC as an irritant who restricts his ability to sell more gas to potential customers further down the gas value chain. And a downstream petrochemical producer who is seeking cheaper gas supplies might also be happy to squeeze out a

---

[15] I have quoted the Executive Summary of the Strategy Note.

middle man, particularly during periods when petrochemicals prices are low. The salient points of the Strategy Note, which raised the following issues relating to NGC's role, are set out verbatim.

Conflicts of Interest. "*In the small Trinidad and Tobago gas market consisting of only three producers and a dozen consumers, NGC's role has become pivotal, encompassing distinct functions ... With such an extensive mandate, the question naturally arises as to whether the country's interests are best served by a single state company involved in so many aspects of the sector, and whether the potential for conflicts of interests is not overwhelming under these circumstances.*"

NGC's involvement in different parts of the energy value chain is by no means unique, and such involvement does not of itself cause conflicts of interest that are insurmountable. *PETRONAS, which (like NGC) was formed in 1974, and is wholly-owned by the Government of Malaysia, had operated* under governments that, by current Western standards, might have been considered to have been strongly interventionist. Under Section 2(1) of the Petroleum Development Act, 1974,[16] *PETRONAS* was vested with the entire oil and gas resources in Malaysia and was entrusted with the responsibility of developing and adding value to these resources. With a strategy of integration, adding value, and globalization, PETRONAS had ventured into the entire spectrum of downstream activities to add value to Malaysia's petroleum resources.

With NGC's interventions upstream, NGC, through wholly-owned subsidiaries, obtained shareholdings in various upstream gas assets: a 15% interest in the Teak, Samaan, and Poui fields acquired from bpTT; 30% and 8.5% interests, respectively, in two blocks comprising the Angostura field, which were acquired when Total left Trinidad and Tobago. NGC also has a 20% share of Trintomar. In practice, all of NGC's sources of gas (including compressed associated gas from the offshore Teak and Poui platforms) were used to reduce the weighted acquisition cost of gas ("WACOG") that served the interests of downstream customers.

---

[16] http://www.commonlii.org/my/legis/consol_act/pda1974257/.

In any event, given the contractual provisions of gas supply and purchase agreements, it is unrealistic to expect that NGC will 'buy' gas from itself rather than from gas producers (and risk liability for take or pay penalties under the contracts with gas producers).

A more difficult issue is whether NGC, as an aggregator, should invest in a sector in which it supplies gas to customers in that sector. During my tenure at NGC, it was an unwritten policy that, in the absence of a strategic imperative, for example, to facilitate the creation of a new industry, NGC would not invest in that downstream sector. There would be issues of transparency. For example, how would an investor in a particular commodity have confidence that the gas supply terms it had contracted with NGC as the wholesaler are comparable with the terms that NGC had reserved for itself or its affiliate? Today, in the event of a gas shortfall, there might be doubts that the gas supply to all companies in the sector had been prorated on identical terms. Consequently, given the confidentiality of contracts and a general lack of transparency, there is a real risk that companies might perceive (rightly or wrongly) that NGC had abused its position.

Subsidies. *"Such conflicts could arise from the fact that policies which are beneficial from the point of view of NGC may not necessarily coincide with the best economic interests of the country. For example, an increase in the number of industrial users of gas may serve the interests of the gas company but could require implicit or explicit subsidies which may not be optimal for the overall economy."*

It is apparent that the WB Team did not know/understand the NGC system. The gas used in the heavy industrial sector has been mainly concentrated in the petrochemical industries and some gas was also used in industries such as iron and steel. Market forces (for example, projections for global supply-demand and prices) determined whether a company would be attracted to invest in a plant in Trinidad to produce additional supplies of a particular product. Prices in the petrochemical sector were market-related, and all prices were negotiated at arms-length.

In a countervailing duty examination that had been brought by producers in the U.S. against Caribbean Ispat with respect to Steel wire rod pricing, I had to explain NGC's pricing structure to U.S. trade officials who visited NGC. After a physical examination of NGC's contracts, the company received a clean bill of health with respect to the challenges to gas pricing for the products and the pier user agreements for shipping those products to the U.S.

The light industrial sector, which represented 2% of NGC's sales and gas volumes, was expected to grow from 12 MMscfd to about 16 MMscfd at the turn of the century. In this sector, all gas prices were negotiated without subsidies, and the gas prices were the highest due to the higher cost of gas distribution in comparison with the large gas users. A two-tier pricing system had evolved, but under a revised policy, it had been proposed to have a unified price in this sector by the year 2000. NGC had offered discounts in the base gas price, as incentives, to new customers who were able to demonstrate that they would be either providing a benefit to the country or assisting in the utilisation of new technology. In that regard, natural gas for air conditioning at the Eric Williams Health Sciences complex, for example, had been targeted.

Internal Conflicts. *"Similarly, within a company with such extensive responsibility, there may exist internally inconsistent interests. For example, the transmission branch of the company has a strong incentive to expand the volume of sales as much as possible, whereas the merchandising branch of the company may want to hold back some supply in order to keep prices higher. Such conflicts will lead to an inefficient allocation of resources."*

Internal conflicts between the two corporate groups were highly unlikely and implausible. Indeed, it went 'against the grain' of NGC's so-called "merchandising" arm "to hold back some supply in order to keep prices higher" in the light of NGC's mandate to monetise gas reserves. In practice, the practical concern for the Business Development Group was whether the existing (and proposed) pipeline configuration had had sufficient redundancy to facilitate and promote additional sales to new downstream customers, as borne out by the Group's strategic goals in 1997:

(a) *"To consolidate Trinidad and Tobago's position as the world's leading exporter of primary petrochemical products - Ammonia & Methanol;*

(b) *To deepen the value added from natural gas through downstream petrochemical development;*

(c) To balance the sales mix through the development of the metals and other heavy industries that utilise natural gas primarily as energy or a reducing agent;

(d) To maximise the use of natural gas within the domestic economy."

The WB Team had clearly got it wrong!

Vertical Integration. *"It is important to unbundle the gas chain, separating the upstream activities from the transmission, distribution and downstream activities. Vertical integration should be restricted as much as possible."*

There are good arguments for greater transparency along the gas chain (discussed later). However, many companies have adopted vertically integrated strategies in order to facilitate better control of processes along the gas value chain. It is a strategy widely adopted globally by several companies in the energy sector.

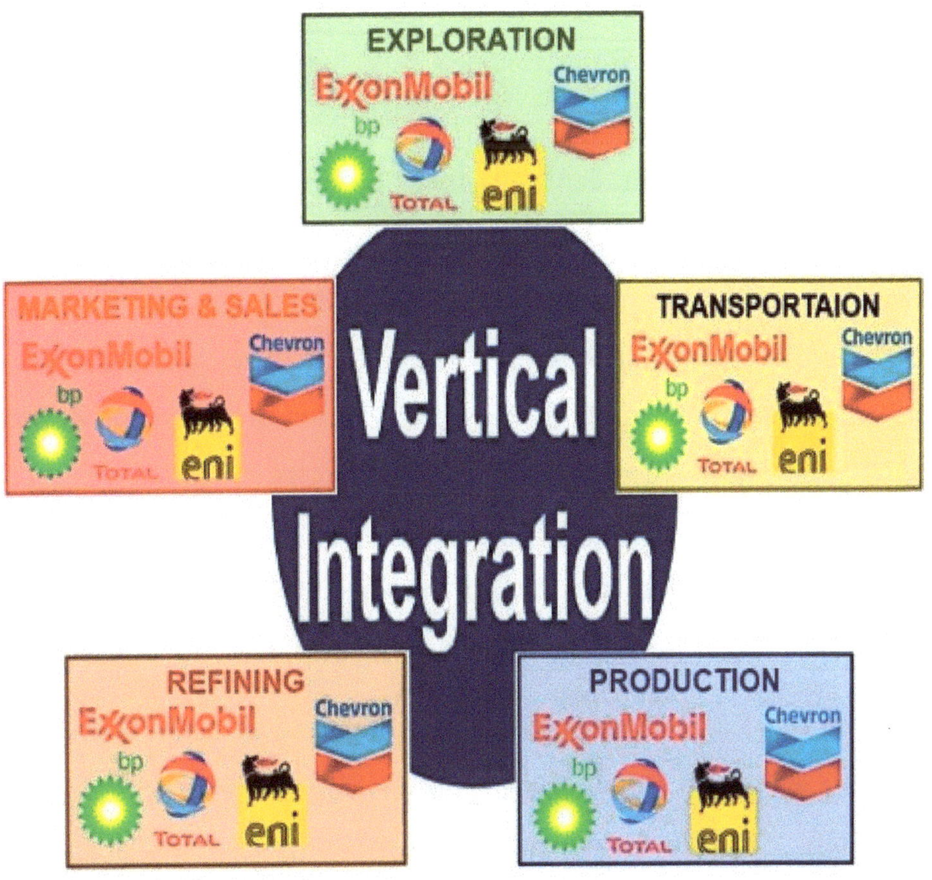

bp is vertically integrated, and bp also owns a convenience store chain that is attached to its gas stations. Shell is active in every area of the oil and gas industry, including exploration, production, refining, transport, distribution and marketing, petrochemicals, power generation, and trading. Petronas, which was under British colonial rule until 31st August 1957, was vertically integrated.

It had appeared to me that the WB Team's recommendation was similarly flawed.

A liberalised Energy Sector. *"While the transmission and distribution activities should be regulated by the government to ensure open access to all producers and consumers, the exploration and production activities as well as the downstream industries should function in a competitive environment."*.

The WB Team's proposed liberalisation of the energy sector involved the following three issues:

(a) Open access to NGC's pipeline infrastructure;

(b) A competitive environment upstream and downstream; and

(c) Regulatory oversight by the Government.

Open Access. The gas market in Trinidad and Tobago differs from those in the U.S. and the U.K., where there is considerable stability, both in market size and price levels. And, with limited available land for industrial development in Trinidad and Tobago and the high initial cost of establishing a transportation and distribution system for natural gas, there is economic (and environmental) justification for ensuring the non-proliferation of cross-country pipelines and one owner of the onshore pipeline system.

In 1997, during my tenure at NGC, the natural gas market had the following market structure (which is materially unchanged for comparative purposes with both the U.S. and U.K. industries):

(a) The local industry had four gas producers, contrasting with the U.S. and U.K. upstream environments with thousands or dozens, respectively, of gas-producing companies;

(b) Approximately 75% of the natural gas produced in Trinidad and Tobago had been utilised for export markets, and the gas prices that were applicable for these markets were determined after arms' length negotiations between NGC and gas producers, on the one hand, and between NGC and downstream users, on the other hand;

(c) Approximately 23% of natural gas was utilised for power generation;

(d) There were about 102 small commercial customers who used about 2% of the total natural gas consumption for cooling and other purposes; and

(e) The natural gas pipeline distribution system did not (and still does not) distribute gas to homes.

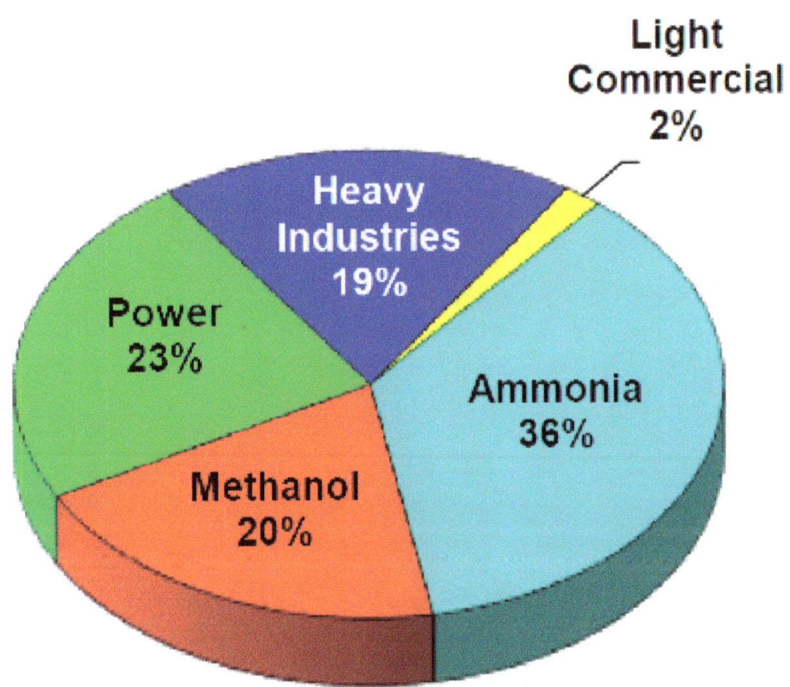

With little gas being consumed in the non-export sector, Trinidad and Tobago had a different political environment to the U.S. and U.K., both of which had (and still have) millions of domestic customers looking to gas-to-gas competition to improve service standards and lower retail prices. Accordingly, the conditions that led to open access to the British Gas pipeline infrastructure or to the regulatory oversight in the U.S. did not then (and still do not) apply in Trinidad and Tobago.

Upstream Competition. In considering what should be done about upstream competition, one must look at the way the industry is organised. In a free market, any upstream producer who can sell gas economically at the lowest prices will be able to secure markets. Without counterbalancing steps, this unconstrained freedom could also frustrate the Government's objective of facilitating diversity among the suppliers because the dominant supplier will always have the flexibility and marginal cost production opportunities that its large infrastructure would provide. Further, in Trinidad and Tobago, which is a developing country with heavy reliance on the energy sector, how would the Government attract new players to explore for petroleum resources if a dominant producer can always beat off the competition for the market?

Since the price of natural gas is the most important variable influencing the natural gas business, it is important for the Government to ensure that there are measures to ensure that it derives maximum economic rent from a depleting asset (which it owns). And, unlike the U.S. and U.K. regimes with liberalised systems, a reduction in gas prices from producers as a result of competition is likely to generate less revenue for the Government unless there is a major restructuring of the whole industry.

The petroleum industry in Trinidad and Tobago has been governed principally by the Petroleum Act, Chap. 62:01 ("the Act") and the Petroleum Regulations, Chap. 62:01 ("the Regulations"). [17] The Act and the Regulations together have established a regulatory framework for the grant of Exploration and Production Licences ("E&P Licences") and Production Sharing Contracts ("PSCs") for the conduct of upstream exploration and production operations. Under the licence regime, blocks had been operated mainly by Amoco (now bpTT), the dominant gas supplier in the ECMA, which had then contained approximately 70 % of the total proven reserves. [18] In 1974, the Government introduced PSCs to increase contractual flexibility and, among other things, gain competitiveness with other countries. Both PSCs and E&P Licences have been issued since then.

With a PSC making provision for cost-recovery, there are still disadvantages suffered by the producer operating under a PSC (the "Contractor") when compared with a company enjoying an E&P licence (a "Licensee"). Thus, in Trinidad and Tobago, where a heavy concentration of the reserves had been owned by Amoco, a Licensee, the fiscal terms for other producers have been considerably less favourable than those of Amoco.

---

[17] The analysis is based upon the regulatory framework at the time of the WB Team's recommendations.
[18] In 2015, it was estimated that bpTT held ~60% of the total gas production and ~ 55% of the proven reserves.

Consequences. A gas producer's ability to market gas is a key factor in reviewing an investment because the timing of the revenue stream is a critical component in determining project economics. In the process of analysing the likelihood of an investment on a discounted cash flow basis, an investment is unlikely to be made unless it is expected that proven reserves could be brought to the market shortly after discovery.

Amoco's strong upstream position has had important consequences for the local gas industry because that gas producer would nearly always have been able to offer gas more cheaply for an equivalent IRR than other gas producers, particularly in respect of greenfield operations. Consequently, without any intervention by the Government, other producers would always be very uncertain about their ability to market any gas that they found.

Interventions. In principle, the problem faced by producers (other than Amoco) could have been removed by either (a) reducing the disparity in the terms between Amoco and the other producers by improving the PSC terms or (b) by managing the situation (at least in part) so that other producers could have access to the market when they have gas. Clearly, wholesale changes in PSC terms would negatively have impacted the Government's revenues (and might be counter to the policy of maximising value). In any event, such changes might not have solved the problem of a disparity because Amoco would still have had the advantage of existing infrastructure, most reserves, and, therefore, ready access to all the market.

Alternatively, the Government might have been able to control the situation by changing Amoco's terms or by asking Amoco to sell down. However, this approach might have involved interference with property rights, which, in the absence of extreme circumstances, might have risked a negative impact on investor confidence generally.

Downstream Competition. This type of competition was irrelevant because NGC's customers were in relation to export industries (75%), power generation (23%), and the light industrial and commercial sector (2%). It was clear, therefore, that the WB Team had made this recommendation without understanding the local context.

Regulation of the Industry. In practice, NGC has been a de facto regulator with two roles:

(a) As an aggregator, NGC diversified upstream gas supplies by entering gas supply contracts with, for example, EOG Resources and BG for the "domestic" market in 1994 and 1996, respectively. Thus, NGC had ensured that the dominant producer did not have a stranglehold on the ability to supply the domestic market.

(b) As a wholesaler, NGC poled upstream gas supplies to make the weighted average cost of gas ("WACOG") attractive for investors in diverse areas such as petrochemicals and steel products.

Managing the market in this manner has provided the flexibility that was appropriate for the size and context of the local industry.

Gas Utilisation. The Government's revenue has had a greater dependence upon energy than in both the U.S. and the U.K., which have more diverse economies. NGC's ability to diversify gas utilisation has mitigated the effects of downturns in any one product, which has been an important consideration for a developing country that has relied heavily upon revenues from natural gas.

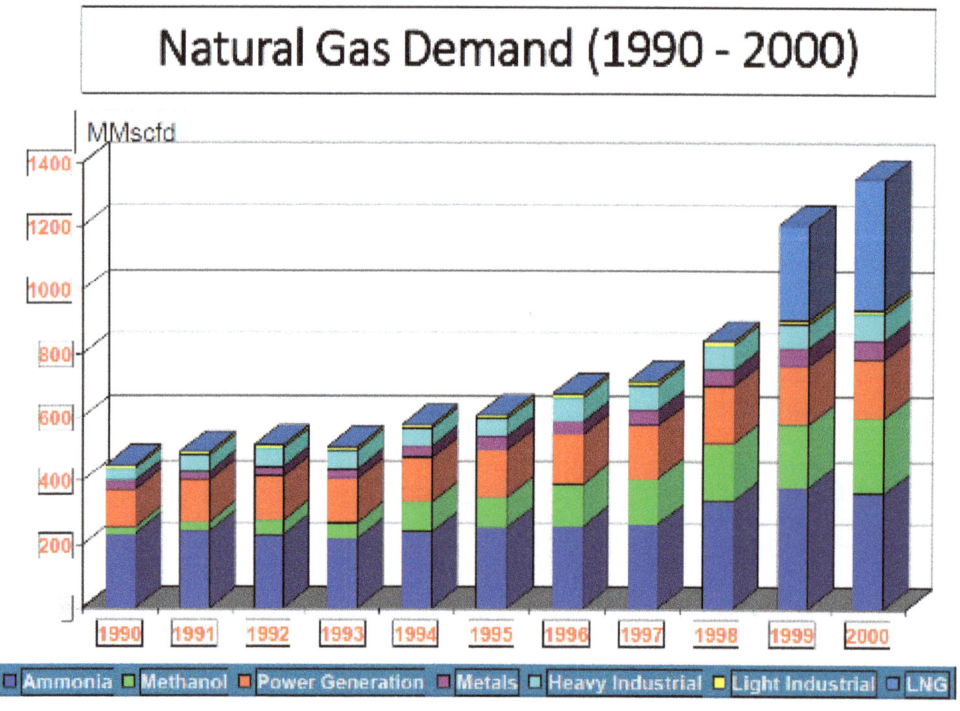

In a truly liberalised energy sector, these strategic initiatives would be compromised because the marketing decisions would be made by the gas producers, who would generally prefer to produce large volumes of gas for LNG production rather than smaller volumes for industries with the prospects of value-added benefits.

Gas Pricing, Efficiency, and Risk. *"The role of NGC as the sole buyer and seller of natural gas is arguably the most controversial. It is claimed that only through the NGC's involvement in the purchasing and selling of gas, can the national interest be defended in the gas business. Evaluation of this argument involves addressing the critical areas of pricing, efficiency and risk."*

Gas Pricing. In 1997, Arthur D. Little ("ADL"), an internationally reputable gas consultant, reviewed NGC's gas pricing policy. ADL established that several gas-rich countries in the Middle East, Far East, and Latin America, with significantly more gas reserves, were potential competitors to Trinidad and Tobago for implementing new gas-based industries. However, the economic and political stability and the commercial terms that could be offered were critical factors in assessing Trinidad and Tobago's competitive position. ADL endorsed NGC's approach to gas pricing to attract new industries (considered later).

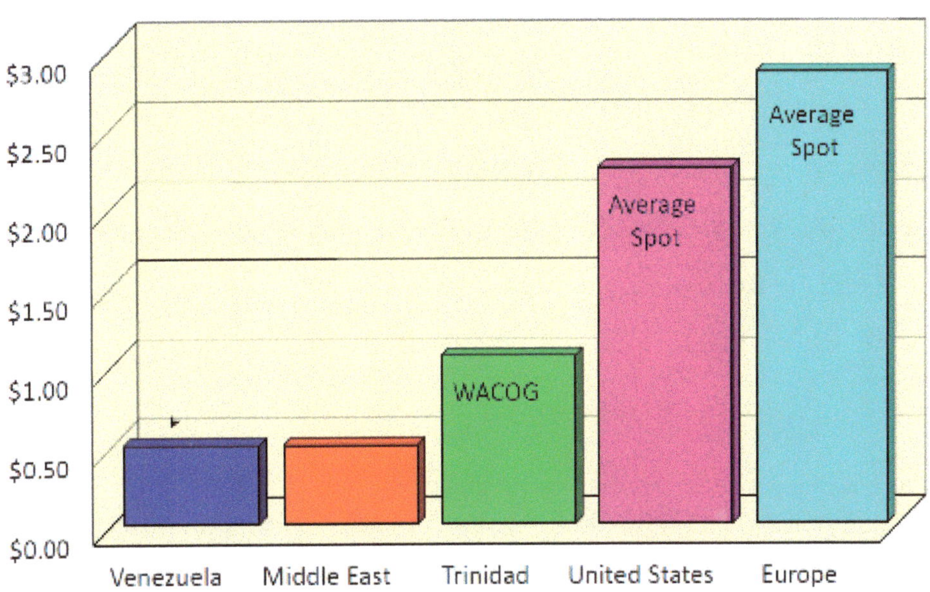

Efficient Taxation System. *"The elimination of NGC's merchant role would stimulate gas-to-gas competition which in turn would have two effects: 1) it might lower the selling price of gas and thus lower upstream tax revenues; 2) it would stimulate economic activity through lower gas prices and transfer rent to downstream industries. The second benefit outweighs the upstream loss of tax revenues especially since the government can recoup these losses by implementing an efficient tax system downstream. [Whilst upstream taxation is] the only real source of tax revenue .. in terms of the future development of the sector, NGC's role as the sole merchandiser could be phased out to be replaced by a more extensive tax system. Such a move would have to be accompanied by a strengthening of the regulatory framework to ensure arm's length negotiations so that producers are unable to transfer profits to untaxable entities abroad.*

Efficiency. The existence of a middleman between the producer and the industrial consumer creates an inefficiency and absorbs part of the margin between sale and purchase price, a benefit that would otherwise be passed on to the domestic consumer or, in the case of gas-based industries, in higher export volumes. The loss of revenue to NGC would be compensated for by increased tax revenue. The fear that gas-to-gas competition would drive out the smaller producers would be unfounded if the government used its contracts and licensing policies effectively. Collusion among producers could be minimized by an appropriate regulatory framework."

"Risk. [S]ome of the industrial consumers express a preference for negotiating directly with the producers of gas despite the risk reduction inherent in the existing commodity-linked prices. One of the benefits of dealing directly with the gas producers is that purchasers have recourse to international law to ensure that the terms of the agreement are upheld; this factor can be important in guaranteeing supply. In any event, the question remains as to whether or not the government should continue absorbing the risk that would otherwise be borne by the buyers and sellers of gas; or in practical terms whether NGC should make efforts in new contracts for passing these risks to other stakeholders".

It is highly questionable whether the WB Team's proposals could have been easily implemented, and many aspects of the proposals do not survive after closer scrutiny:

(a) **Shareholder Perspectives**. In much the same way that private sector participants along the gas chain seek to maximise value for their shareholders, NGC has a duty to maximise value for the Government (and ultimately, the citizens of Trinidad and Tobago).

(b) **Gas – to – Gas Competition**. In practice, would there be gas-to-gas competition upstream where there was a dominant producer? In any event, a change in the *status quo* would have had to be accompanied by the re-structuring of the entire natural gas industry and the operating legislative framework to ensure that the change did not discourage new producers, drive out smaller producers or negatively impact the Government's revenue.

(c) **Taxation**. It would not be prudent to assume that loss in tax revenue either from the upstream or from NGC would be offset by an undefined (and highly speculative) taxation system:

  i. In a developing country with limited public-service resources, it is very risky for the State to revamp the entire system to rely upon administrative practices.

  ii. Even developed countries with sophisticated taxation systems have been victims of transfer pricing, where multinational corporations, using very complex legal transactions involving subsidiaries and related companies, have been able to shift profits that are due to a host country into extraterritorial tax havens.

  iii. NGC's merchant roles (as aggregator and wholesaler), without open access to its pipeline infrastructure, have provided an institutional structure that has reduced the risk of parties colluding to share a portion of the State's economic rent outside of Trinidad and Tobago's jurisdiction.

(d) **Regulation**. It is doubtful that a regulated system would have been appropriate for the local context at the time, which contained few producers and consumers:

  i. A system like OFGAS would have been expensive and burdensome to implement.

  ii. Arms-length negotiations with quick decision-making and flexibility had given Trinidad and Tobago a major advantage in monetising gas reserves.

However, there could be justifiable changes to the *status quo* in **certain** circumstances:

  i. A regulated system might become feasible if the industry were to reach a stage where there were sufficient producers and consumers for a free competitive market, in which case NGC's role as buyer and seller of gas could be phased out and replaced by an extensive regulatory system that ensured arm's length negotiations.

ii.   It would facilitate more efficiency if NGC kept separate accounts for its transmission and its merchant and other functions [19] because separate accounting might result in more efficient services for all participants in the industry.

(e) **Risks**. The World Bank team had questioned whether NGC should continue to absorb "the risk that would otherwise be borne by the buyers and sellers of gas," whereas, in contrast, ADL had recommended that NGC should retain market-related pricing:

i.   Market-related pricing had been developed in the energy sector in recognition of the cyclical nature of many petrochemical products, and it had been the locomotive for NGC's success in attracting natural gas-based industries.

ii.   The market-related pricing mechanism follows the format of a base price tied to an ammonia or methanol price, with the base price fluctuating according to market conditions. The mechanism limited NGC's upside profits with a "cap," but it protected NGC with a "floor" gas price in depressed market conditions.

iii.   The parties generally would agree on the pricing terms after arms-length negotiations (often in conjunction with input from the project's financiers).

A Recurring Theme. Arguments that NGC, the middleman, should be removed, or its margins for the benefit of the State, should be reduced often recur in the following circumstances:

(a) As attempted leverage to obtain price reductions when cheaper gas was available elsewhere;

(b) When a petrochemical producer wished to maintain the original return on his investment because older plants had become less efficient (and less profitable); or

(c) As an attempt, when market prices were low, to renegotiate the price risk agreed upfront.

---

[19] Perhaps this is already being done (given the existence of NGC Pipeline Company Limited, which ought to have a Board of directors).

Upstream Risks. NGC initially took the supply risk in contracts with gas producers by agreeing to a contractual term that a failure by gas producers to supply gas to NGC was a "force majeure" event. This contractual provision was passed on to downstream buyers. Today, ideally, gas supply contracts should be negotiated with warranty (reserve-backed) contracts, which was the approach taken during the negotiations with Amoco in relation to the 1996 Gas Supply Contract.

Gas Pricing Transparency. In the natural gas industry, where the environment is dynamic and the applicable gas price is determined through arms-length negotiations, it is to be expected that there would be variations in some of the commercial terms of gas contracts. The price that a gas producer could offer would depend upon his circumstances at the snapshot in time. In the same vein, changing market conditions will impact the gas price or risk that a potential petrochemical investor would be able to accept for a project at different times in the same industry. Whilst the rationale for the lack of uniformity in gas pricing has not always been understood (and has led to criticism), Trinidad and Tobago had been able to maintain its competitive edge in attracting upstream and downstream investments by having flexibility in gas price negotiations.

As to transparency, arguably, contractual matters should remain within the province of the members of the Board who have fiduciary duties to the company (and, by extension, act as trustees for the people of Trinidad and Tobago). However, in the absence of transparency, how can the public be satisfied if members of a State Board are often unsuitable and/or appointed at the whim of politicians? Perhaps Bereau JJ provides a solution: [20]

> *"It is a matter of great public notoriety that directorships in state enterprises in Trinidad and Tobago are much more a question of political patronage and cronyism, than it is about competence and in which the lines between self interest and the public interest can become blurred. In those circumstances, it is hardly likely that wrong doing by directors will be ferreted out before a change of government."*

---

[20] Per Bereaux, J.A., Kenneth Julien v Evolving Tecknologies and Enterprise Development Ltd, Judgement, 14th August 2014 (at paragraph 26).

Accordingly, regardless of any lack of transparency, NGC's directors will always be liable for any breaches of their duties to NGC because if there has been wrongdoing by the directors, the lapse of time before any new directors are appointed would not be relevant for the purposes of assessing the time period under the Limitation Act, should litigation be commenced later.[21]

The Poten Report. In September 2015, Poten and Partners ("Poten") prepared a Trinidad and Tobago Gas Master Plan for the Ministry of Energy and Energy Industries,[22] which considered (among other things) NGC's role. An overview of Poten's views is provided below.

"1.12.4.2 Future Downstream Contracts, Sector Structure & Role of NGC

It is clear that the original drivers behind NGC's adoption of an intermediary / wholesaler role in the T&T gas sector no longer exist. The industry has moved from a growth / development phase of plentiful gas supply where a market-making function was required, into maturity, and is now facing the prospect of a situation under which gas supply is highly unlikely to be sufficient to fully meet demand going forward.

Indeed it can be argued that the role of NGC is complicating the operation of the sector as it struggles to match supply with demand and that for large buyers this would be better and more economically efficiently handled by large buyers and sellers interacting directly. However, from GORTT's point of view the key factor that must be considered is the significant economic rent captured by NGC in the midstream and ultimately distributed back to GORTT as a dividend. If the wholesale margin was passed back to upstream then GORTT would have to share the upside with the upstream suppliers as per the terms of the various upstream agreements. With this in mind, Poten's view is that the uncertain benefits associated with a significant restructuring of NGC's role as wholesaler / transporter are unlikely to be justified [by] the challenges associated with maintaining existing GORTT take levels under a new structure (e.g. by imposing new taxes), and the time and cost associated with implementing what would undoubtedly be a major restructuring exercise.

---

[21] Julien and others (Appellants) v Evolving Tecknologies and Enterprise Development Company Limited (Respondent), Privy Council Appeal No 0065 of 2015, [2018] UKPC 2.

[22] The Final Report can be obtained from the following link provided through the Ministry's website: https://www.energy.gov.tt/wp-content/uploads/2020/01/Trinidad-and-Tobago-gas-master-plan-2015.pdf.

As such we do not believe that allowing the bypass of NGC, unbundling NGC's transportation activities, or fully liberalising the sector will be optimal routes for GORTT to follow.

Rather than maintaining the status quo, Poten's view is that, on expiry of the existing LNG contracts, GORTT should seek to expand NGC's wholesale role to include supply to ALNG, i.e. NGC would buy gas from upstream and sell it to or toll it through ALNG. Although this is very much an interventionist approach, Poten's view is that it is likely to maximise GORTT's overall take from the sector, without compromising the ability of the sector to provide more attractive prices to upstream in order to support new developments. In addition, this option would allow NGC to manage gas supply to the whole downstream sector, whereas at the moment it has limited control of how much gas is supplied to LNG. This is of particular relevance in a gas shortfall situation.

Poten's view is also that NGC's business should be refocused on its core wholesale & transportation activities, i.e. its other non-core assets should be divested, potentially either to other existing or new GORTT entities, or to new publicly-owned vehicles. There is no obvious reason as to why NGC is the best undertaker of its non-core roles, such as sector business development, or the best holder of its noncore assets, e.g. upstream production. In particular, these roles create potential conflicts of interest for NGC's core role. This would not preclude the Government continuing to hold these assets; indeed it is recognised that positions in upstream and downstream assets provide valuable information.

Extending NGC's wholesale role will also increase the oversight required of NGC's activities by GORTT to ensure that it is acting in the broadest interests of GORTT rather than its own more limited perspective. In parallel with expanding NGC's wholesale role to include LNG, Poten recommends that the centrally planned, allocative approach to future downstream gas contracting is adopted. For the same reasons put forward for the future role of NGC, Poten does not believe that adopting the market-based approach will be in the best interest of T&T.

Under the two centrally-planned approaches there are clear attractions to the tendering option which would potentially provide a transparent and fair price discover process.

However, our view is that the obstacles to implementing this option (establishing tender parameters between different commodity producers and between plants with different contract expiry dates) will be very difficult to overcome in practice. This leaves the approach under which GORTT determines the downstream consumers that will receive gas as the only viable option. In terms of implementation, there will need to be an assessment made by GORTT/MEEA/NGC as to how much gas will be allocated to the key consuming sectors, e.g. LNG, ammonia, methanol and steel, as it is unlikely that there will be sufficient gas to fully satisfy demand. This analysis will rely on projections of expected future value to GORTT, which in turn will be highly dependent on projections of future commodity prices, which are inherently volatile and unpredictable. As such, although for example LNG may be projected to provide the highest value to T&T, GORTT may determine that it is in its interest to maintain a broader downstream portfolio in order to insulate itself from future global market changes, i.e. rather than fully filling LNG demand and shutting down various ammonia / methanol plants, GORTT may decide to reduce supply to LNG somewhat in order to maintain supply to ammonia / methanol.

Within the determination of how much gas to be supplied to each sector GORTT/MEEA/NGC will need to decide which plants should receive an allocation of gas and which, if necessary, should be shut down. While it will be a difficult decision to shut down a downstream plant, this will inevitably need to happen over time. If, for example, there is only sufficient gas to keep 50% of T&T's methanol capacity operational it will be far better from an economic perspective to shut down half of the plants and keep the remainder operating at full capacity, rather than keeping all of the plants running at half capacity but with full running costs. It should also be noted that although plants can be mothballed for a period of time and then brought back into operation if gas subsequently becomes available, in practice it will be costly to maintain plants in a mothballed state, keep staff etc.

Based on NGC's contracted upstream gas supply, an assessment will also need to be made for how long NGC can provide downstream gas allocations. Although all of the downstream plants in question will have been fully amortised by the time that their existing gas supply contracts expire, buyers will need some certainty over future gas supply if they are to make investments which may be needed to prolong the life of the plant. With its expanded wholesale role, experience of managing its existing downstream sales portfolio and share of GORTT's overall gas sector knowledge and expertise, NGC should be well-placed to provide the necessary analysis and recommendations to GORTT/MEEA on downstream gas allocations. However, there should be strict guidelines in place about how allocations should be made, i.e. maximising GORTT take from its gas resources, and GORTT/MEEA should have the ultimate decision-making power regarding any new gas allocations. GORTT/MEEA/NGC will also need to consider the potential allocation of gas to any new industries in parallel with its analysis of allocations to existing users."

Scope for Optimism. I generally agree with most of the Poten analysis, which in many respects, endorses some of my criticisms of the WB Team's recommendations in 1997, particularly those that concern the liberalisation of the local energy sector. However, while, in the current context of gas shortfalls there is scope for doubting NGC's continued role as a middleman, the situation could change because the longevity of the energy sector remains an open question until the full scope of the Nation's gas resources has been determined. There are also still significant prospects for cross-border gas from Venezuela because of U.S. policy changes.

Conflicts. Although no specifics were cited, perhaps, the transfer of NGC's upstream assets to another State entity might reduce (or remove) potential conflicts since the directors of the other entity would be under a duty to secure, for example, the best upstream commercial terms for gas sales to NGC. However, it will be important to evaluate, upfront, at the very least, the costs and human resource requirements that would be required before attempting to implement the proposal.

Regulatory Body. Extending NGC's wholesale role will increase the oversight required for NGC's activities, which should be pursued subject to a cost-benefit analysis. As an observation, however, in the short term, it is far from clear that the public sector has the resources (or expertise) to ensure that **<u>ALL</u>** actors in the local energy sector comply with regulations and do not engage in, for example, anti-competitive behaviour. Ultimately, the system could evolve to include a regulatory body to determine the tariffs. However, the industry would have to develop to a state where the implementation of such a regulatory body had become justifiable at the time in the light of the requirements for the appropriate quantity of capital and human resources.

Expanded Wholesaler. I strongly agree with Poten's view that the Government should expand NGC's wholesale role to include supplies for LNG. In principle, if gas producers have a choice, in the absence of contractual commitments to do otherwise, they will prefer to give priority to gas supplies for LNG in order to protect their significant investments at Atlantic LNG. In the same vein it will be commercially prudent for the gas producers to channel gas supplies to LNG Expansion Trains where the returns are more likely to be higher than in respect of Train 1.

However, I believe that the Government should take an interventionist approach and obtain greater control **sooner rather than later** and, in any event, before all the LNG contracts expire[23]:

(a) The effects of the war in Ukraine are unpredictable and might have geopolitical consequences that require State involvement in decisions about gas utilisation;

(b) It appears that the marketing arrangements with downstream entities of the LNG offtakers have resulted in massive value capture falling outside the Trinidad and Tobago tax net;

(c) In an environment where there is a shortfall in gas supplies, LNG priority should be constantly reviewed because it consumes >50% of Trinidad and Tobago's gas production;

(d) As a corollary, the global supply/demand for natural gas is in flux, and, at some stage, LNG might provide significantly lower returns than either ammonia or methanol; and

(e) Future cooperation for example, with respect to the implementation of "green" hydrogen projects to address climate change, are likely to have more traction with producers when the Government has leverage before the Train 4 contract expires in about five years.

Offtaker from Train 4. The benefits of placing the offtaker function out of NGC into a State entity without links to NGC (as Poten suggests) are unclear. Why create an independent entity when an NGC subsidiary, as the LNG offtaker, is likely to facilitate better communication of

---

[23] Shell and bp cooperation will eliminate arguments about a *force majeure* event and enhance their reputations. This is considered later in the final chapter under proposals relating to the Heads of Agreement with bp and Shell.

market information, internal human resource development and more efficient decision-making in the long term? A summary of the Poten recommendations is provided in the Table below.

Table 1-12   Options for Gas Sector Development – Role of NGC

Option	Rationale	Implementation Requirements	Comments / Issues
No Change	• The existing situation is better than all the alternatives	• Business as usual	• NGC will increasingly be caught up in conflicts of interest as gas allocation decisions occur more frequently • NGC margin will be subject to erosion • Volume mismatch risk remains with NGC
NGC becomes single buyer for all gas in T&T	• Efficient route for GORTT to extract value from LNG • Allows NGC to manage supply allocation to the whole sector • Would allow NGC to offer blended prices to suppliers	• As LNG contracts expire NGC incorporates supply to LNG into its wholesale portfolio	• NGC extends monopsonist powers to whole sector • Sector will lack transparency • Potentially increases NGC volume risk • Appetite of some upstream suppliers to accept basket pricing uncertain • Incumbents will likely oppose as existing LNG arrangements have generated substantial value for them
NGC business refocused on wholesaling and transmission	• Removes potential upstream conflicts of interest • Focuses NGC business on core skills	• Divestment of non-core assets (e.g. upstream assets)	• No obvious reason as to why NGC is the best owner of upstream assets • GORTT would have to reallocate divested assets • Could be combined with the role as a single buyer
Allow bypass of NGC by large buyers for new supply	• Increased transparency • Takes volume risk away from NGC • NGC able to aggregate supply from small suppliers if this service is required	• Would require transportation separation & tariff structure development • There would need to be DMO (or similar) on suppliers (~10%) to cover sales to power/steel etc.	• How to ensure that the available gas gets sold to the party willing to pay the most in a shortfall situation? Tender? • NGC presently extracts significant rent from the gas value chain for T&T – how to ensure this continues? Midstream taxation? • May result in NGC stagnation - left with lower-priced contracts in its portfolio.
Transportation services unbundled	• Would result in greater sector transparency	• Separation of transportation and gas supply functions of NGC • Tariff structure development • Regulatory oversight	• Where should the regulatory function sit? MEEA? • Would need to develop Institutional capacity of MEEA
Fully liberalised market	• Removes need for intermediaries	• Breakup of NGC - becomes transportation provider • Open access on the transportation system • Would require DMO for power/steel	• T&T market is not sufficiently deep or liquid to support this option • Not clear how to ensure that all such transactions are arm's length • Opportunity for shifting value along the chain and possibly offshore

Chapter 9: The Future Challenges

Overview. The sins of the past, like chickens, have come home to roost: significant value-added opportunities have been lost, gas producers control LNG marketing, and the gas reserves are almost depleted. From a long-term perspective, the impact of gas prices in the U.S. is not encouraging. And, how will any future investors assess the 'country risk' of Trinidad and Tobago? Thus, the Government now has the extremely difficult task of resolving immediate economic challenges while attempting to preserve options to sustain the energy sector.

The LNG Trains are depleting significant quantities of gas reserves. Accordingly, would it be feasible for the Government to use the Heads of Agreement with bp and Shell to control how the limited natural gas resources are now utilised? Should there be natural gas allocations between the plants at Point Lisas and Point Fortin? Should decisions on any gas allocations for petrochemical plants be based upon their value-added economic potential? One might speculate but with a caveat that the best solutions require knowledge within the province of the Government. Under all scenarios, the following objectives to increase locally available natural gas reserves are relevant:

(a) An evaluation of the terms in PSCs that would encourage exploration and production or incremental development in an increasingly competitive regional environment.

(b) The implementation of cross-border gas initiatives with neighbouring Venezuela.

Shale Gas. Trinidad and Tobago has had a competitive advantage as an exporter of LNG and gas-based petrochemicals (ammonia/urea, methanol) because of its low cost of gas and the proximity to the world's largest market, the U.S. However, these competitive advantages have eroded over time due to the advent of shale gas and coal bed methane in the U.S. market.

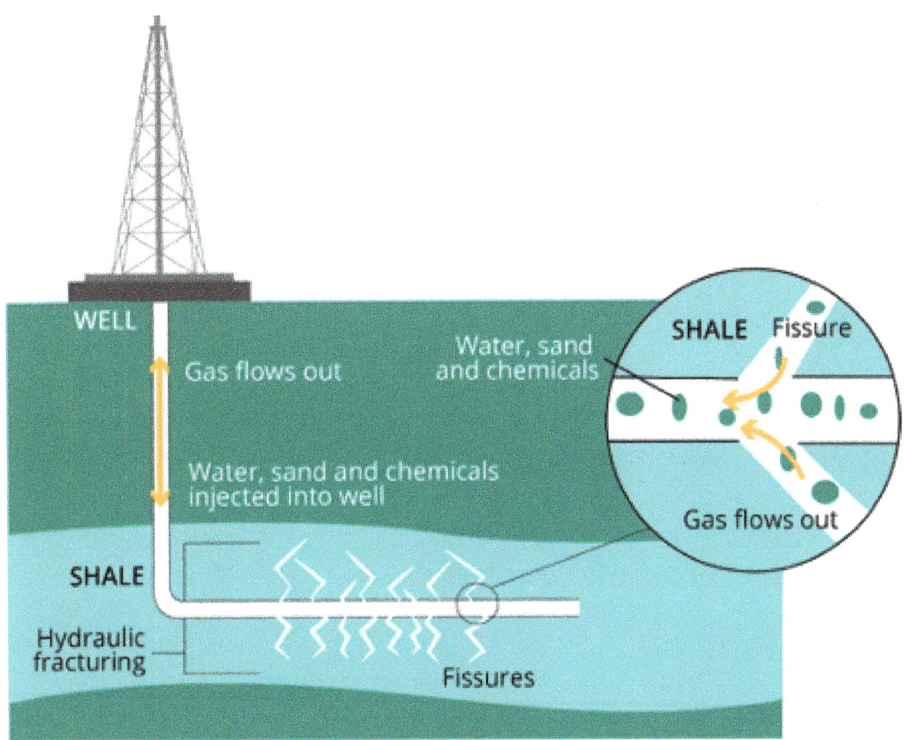

The U.S. Energy Information Administration (EIA) has estimated that U.S. dry shale gas production was about 26.3 Tcf, and equal to about 79% of the total U.S. dry natural gas production in 2020[1]. Shale is now the largest source of U.S. gas production and according to the EIA, the unproved technically recoverable U.S. shale gas resource is estimated at 482 Tcf.[2] Accordingly, in the future, not only will any incremental gas supply from Trinidad and Tobago's reserves be more expensive to develop, even with climate change reductions due to climate change policies, the U.S. market will become even more saturated with gas, bolstered by the rapid growth of shale gas, which can be developed at a relatively low cost.

---

[1] https://www.eia.gov/tools/faqs/faq.php?id=907&t=8

[2] U.S. Energy Information Administration, "Annual Energy Outlook 2012," Table 14, Unproved technically recoverable resource assumptions by basin, page 57.

As gas production increased in the U.S., prices have fallen, and U.S. demand for imported LNG has declined. Also, according to the EIA,[3] the U.S. LNG gas export capacity has grown rapidly since the lower 48 states first began exporting LNG in February 2016. In 2019, the U.S. became the world's third-largest LNG exporter, behind Australia and Qatar.

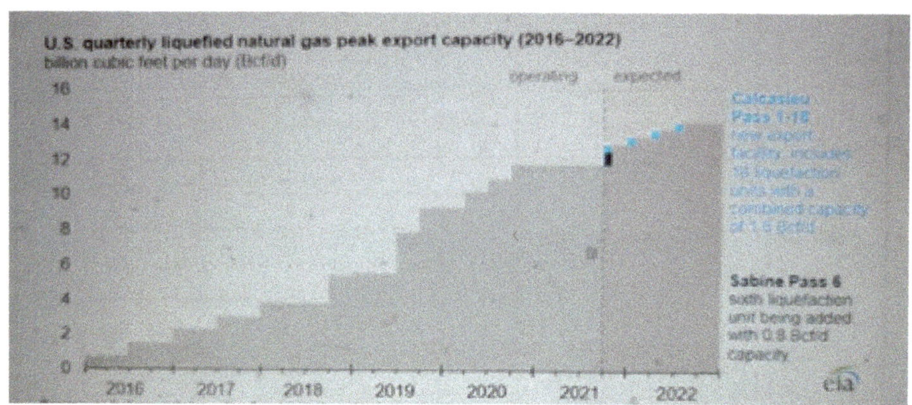

The increase in LNG terminals in the U.S. will also reduce natural gas prices in the U.S.

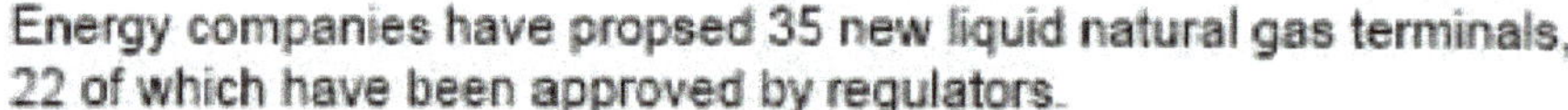

Energy companies have propsed 35 new liquid natural gas terminals, 22 of which have been approved by regulators.

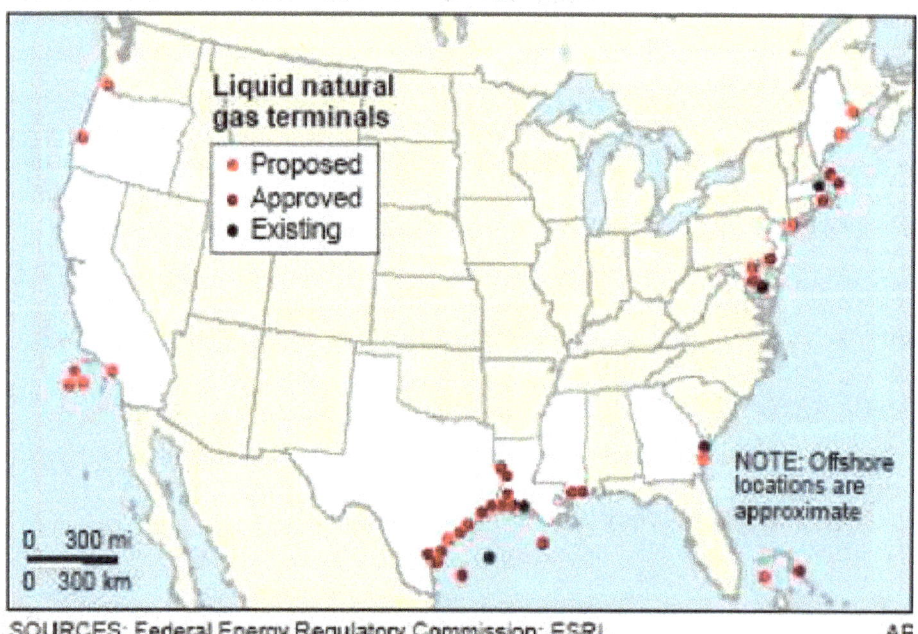

---

[3] U.S. Energy Information Administration, 'U.S. liquefied natural gas export capacity will be world's largest by end of 2022', 9th December 2021, https://www.eia.gov/todayinenergy/detail.php?id=50598.

In any event, over time, Trinidad and Tobago's natural gas products will increasingly compete for market share against products from other supplier countries. As these products are progressively pushed out of the North American market, they will have to travel further to reach new markets, which will add to logistical costs and reduce netbacks to Trinidad and Tobago.

Country Risk. Arguably, Trinidad and Tobago's stable political climate (which is a component of its "country risk") has been the pre-condition that has tipped the balance in the Nation's favour when investors (and their financiers) conduct due diligence on a new or incremental project.

Thus, a prospective investor will be discouraged from investing in any country that is perceived to be politically or socially unstable, even if the country can offer internationally competitive financial terms for the investment. These country risk factors have always given the Nation a competitive edge over neighbouring Venezuela and the Middle East, both of which have had significantly larger quantities of natural gas reserves and have been able to offer lower gas prices to the world.

As an illustration, when assessing Trinidad and Tobago's political/country risk in relation to Train 4, BG's assessment in March 2002 took account of the following considerations:

> *"Although the two major parties generally represent different ethnic groups and regions (PNM – urban/afro, UNC – rural/Indian), there is little difference in their platforms. Both parties agree on many areas of policy and consequently major swings in policy, particularly energy policy, are unlikely regardless of which party is in power. However, while the investment climate is still considered to be good, a protracted period of political uncertainty will impact negatively on this. If fresh elections again fail to resolve the deadlock, the delay to policy-making decisions, especially in the oil and gas sector, could begin to affect prospects for investment. …… Low ammonia and methanol prices could raise pressure on BG to reduce its selling prices to NGC to match BP's but the Government has steadfastly upheld the sanctity of contracts. Unemployment has remained high as a result of 'jobless growth' in the oil and gas sector. This may be a cause of concern for the future if not tackled by the Government, resulting in increasing personal security problems and increased drug trafficking with associated increases in corruption."*

There are also illusions to concerns about local ethnic issues when, in April 2012, Robert Riley, former chairman of bpTT, gave the feature address at the Trinidad and Tobago Chamber of Commerce's annual meeting business luncheon at the Hyatt Regency (Trinidad) hotel in Port of Spain. He cautioned that *"race issues have been with us for years and have been impacting how the country moved forward"*:[4]

> *"We have not worked hard enough as a nation on developing a vision that is shared and owned by a critical mass of our leaders regardless of race, political or other persuasions that has powerfully, despite our complexities and differences, united us under a common purpose and banner as a nation."*

Since 2012, arguably, lack of accountability, partisanship, and cronyism have worsened in public affairs, including the appointments of the members of Boards in State institutions. If these negative perceptions of politicians are unchecked, many citizens with integrity might refuse public service

---

[4] 'Race issues 'hampering progress' by Curtis Rampersad, *The Trinidad Express*, April 25, 2012 http://www.trinidadexpress.com/business/Race_issues___hampering__progress_-148989305.html (now available in Starbroek News https://www.stabroeknews.com/2012/04/25/news/guyana/race-issues-hampering-progress-in-tt-%E2%80%93-energy-expert/).

to avoid their reputations being tarnished by association. Even worse, if politicians promote a culture where citizens are scared to challenge their decisions, the society can spiral downwards until absolute loyalty, not competence, provides the only rationale for State appointments. It is a slippery slope that could deter future investors and ultimately lead to a "failed State."

Cronyism. The negative impact of cronyism on investment is reflected by the following assessment of Venezuela in 2003 by a former colleague with local Venezuelan insights:

> *"The experienced technocrats in middle/senior management have gradually been replaced over the period that Chavez has been in office by political appointees – effectively giving the Ministry of Energy and Mines greater control of policy-making. Generally, the implications of a weak PDVSA are that there are likely to be more delays as a result of indecision, obstructing decisions that are politically motivated, and the need to involve the Government machinery. At the moment, PDVSA is having difficulty obtaining sufficient funding to complete its exploration programme in the Plataforma Deltana. …. Chavez has not been successful in improving the institutions required for sustainable economic growth, nor the short-term macroeconomic performance. When coupled with the Chavez autocratic style of government, the risk of social unrest is increasing and the difficulty of doing business in Venezuela may therefore be increasing. While PDVSA functioned as a "State within a State", this did not deter investment or delay activity in the oil and gas sector. However, PDVSA has almost lost that role. The possibilities for further delays may therefore have increased."*

The Heads of Agreement with Atlantic LNG. Train 1 is reported to be "dead"; the Train 2 contract ended in 2022,[5] and the Train 3 and Train 4 contracts will come to an end in 2023 and 2027, respectively. The gas reserves are finite so how does one obtain the best value from this depleting resource? Since all the LNG gas contracts will eventually end, we should be reminded of the economic decisions that were taken by Shell and Texaco to leave the operations at Point Fortin and Point-à-Pierre, respectively.

---

[5] I believe that it was renewed (but it has not been confirmed) for a period of two years.

Under the terms of the Heads of Agreement, there might be an opportunity for the Government, with bp and Shell cooperation, to adopt principles to control how gas is now utilised:

(a) Should Shell and bp forego their interests in Train 1 to NGC?

(b) Should 3rd parties be permitted to utilise all available offshore infrastructure (including the LNG Facilities to process gas for the manufacture of LNG) on terms to be determined by the Government? This approach will optimise infrastructure and improve the economics for marginal developments. It was adopted in the U.K. in respect of North Sea operations.

(c) Should the Government determine whether gas is utilised for LNG or for petrochemicals?

(d) Should the Government ensure that-

    i.   LNG is sold in markets that provide the best returns to Trinidad and Tobago; and

    ii.   The Government has audit rights in respect of the sales in any LNG markets?

Cross-Border Gas Initiatives. The Loran–Manatee gas field, the Manakin–Coquina gas field, and the Kapok–Dorado gas field span the marine border with Venezuela and the ECMA. There are also cross-border fields near infrastructure on the western side of Trinidad.

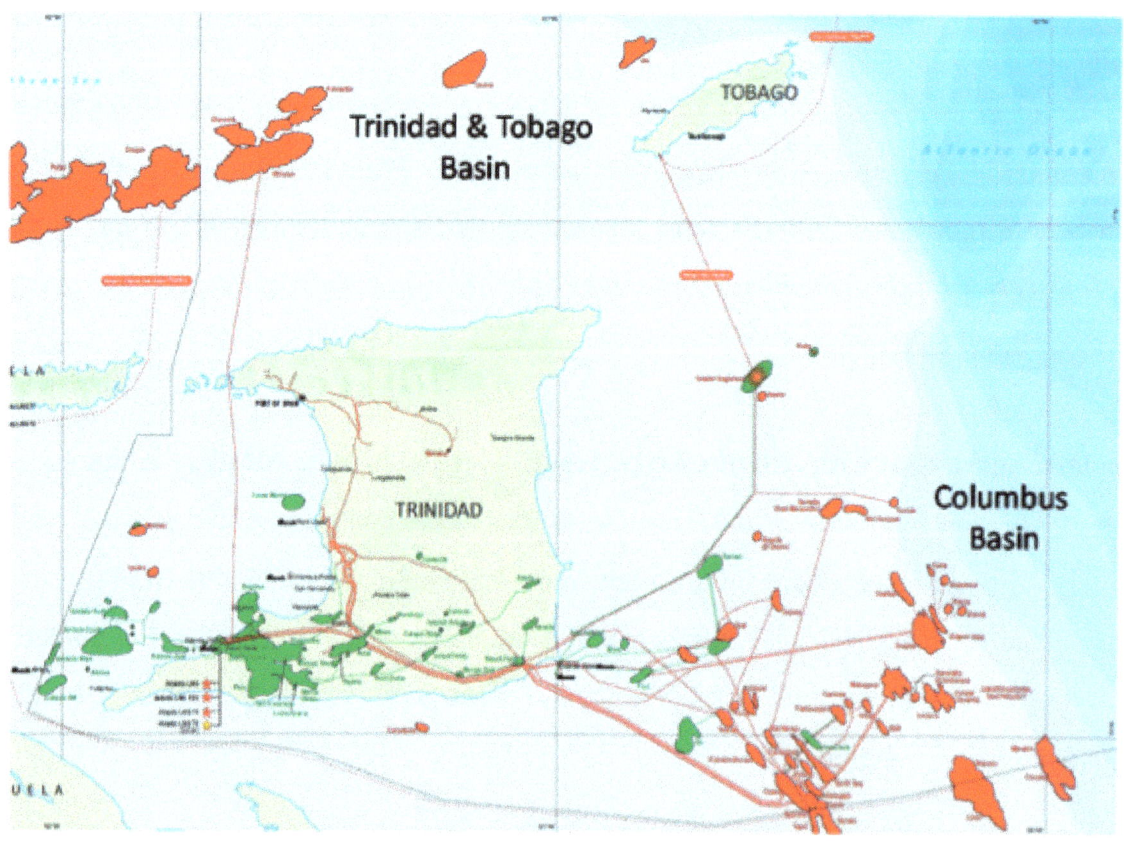

The Kapok-Dorado field has been in production by bpTT, the operator, and by far, the largest field is the Manatee/Loran field. Loran (known internally as the Manatee prospect in Block 6 in Trinidad) was the highest-ranked structure (on a technical basis) in BG's local prospect inventory. In the light of BG's strategic interest in the Loran Field, during the Train 4 negotiations with the Government, I wrote the Chairman of the Government's negotiating team in the following terms:

"Dr. Julien

Following our conversation, I thought it might be useful to jot down some points on the cross-border issues that might help to secure our objectives. As you know, BG & Chevron Texaco - the ECMA partnership - have an interest in Block 6d, which is part of the Trinidad/Venezuela cross cross-border Manatee/Loran field complex. It is estimated that the field contains 5.2 Tcf, 25% of which is believed to be in Trinidad and Tobago's waters. PDVSA have plans for development further south in Venezuelan waters, and as these plans advance, Venezuela could have a less strategic interest in pursuing initiatives with Trinidad. If Venezuela lost interest in initiatives with Trinidad, the cross-border reserves could be stranded for many years, perhaps even indefinitely. Accordingly, with BG's current reserve position in T&T and GORTT's desire to expand the gas resource base, both GORTT and the ECMA partnership have an interest in progressing Treaty negotiations rapidly with Venezuela for the exploration and possible joint development of the relevant cross-border fields ... ."

More recently, in October 2021, while addressing the Senate, the Minister of Energy revealed that major strides had been made and that Trinidad and Tobago's perseverance with the Loran Manatee field had paved the way for the largest gas production deal in decades:[6]

"I am happy to announce here today that last week, the Cabinet of Trinidad and Tobago took a decision and gave us a green light to go ahead with the Manatee contract.....Loran Manatee is a gas field that straddles Trinidad and Tobago and Venezuela. We've heard

---

[6] Peter Christopher, 'Young: Govt reaches Manatee deal with Shell', *The Guardian*, 18th October 2021; https://guardian.co.tt/business/young-govt-reaches-manatee-deal-with-shell-6.2.1401553.339e9f7a25.

The cross-border gas fields would provide significant incremental supply to Trinidad and Tobago, particularly if **all** of the cross-border gas were to be monetised in Trinidad and Tobago.

Geopolitical Considerations. It has been reported that, in 2019, Venezuela offered "sweeter fiscal terms to Russia's state-controlled Rosneft" [7] for the Patao-Mejillones gas fields, which are well-positioned to supply gas to Trinidad and Tobago. Rosneft, I believe, is the sole owner and operator of the Patao-Mejillones gas venture under Venezuela's 2001 gas law and had a concession contract signed in Moscow in 2017. The Patao and Mejillones gas fields form part of PDVSA's Mariscal Sucre offshore gas complex, which also includes the Dragon and Rio Caribe fields. Mariscal Sucre's combined gas reserves total over 14.3 Tcf.

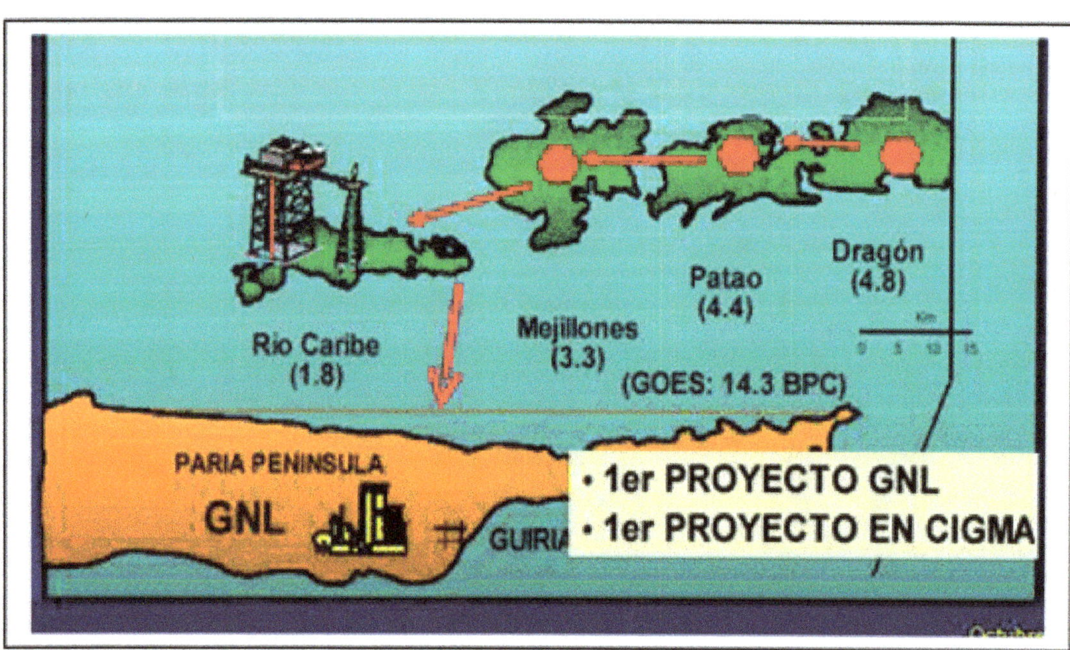

---

[7] Energy Analytics Institute, 'Rosneft Cool On Venezuela's Sweeter Gas Terms, For Now', 6th November 2019; Rosneft Cool On Venezuela's Sweeter Gas Terms, For Now | Energy Analytics Institute (EAI) (energy-analytics-institute.org).

Shell is a party to the stalled 2018 agreement with PDVSA and NGC to develop the Dragon Field and a flow line into its Hibiscus platform off Trinidad. This project did not get off the ground due to escalating U.S. sanctions aimed at forcing President Nicolas Maduro to step down. The Dragon Field is also expected to be more challenging to develop than Loran-Manatee because it requires a separate agreement between PDVSA and Shell. However, there are encouraging signals that U.S. sanctions will be lifted[8] (which will clearly benefit Trinidad and Tobago).

Cross-Border Security of Supply. In any cross-border relationship, there are long-term considerations that ought not to be overlooked. Even before the current Russian-Ukraine conflict, an earlier gas dispute between Russia and Ukraine in January 2009 had led to a supply cutoff to consumers in Eastern Europe, unprecedented in its severity.

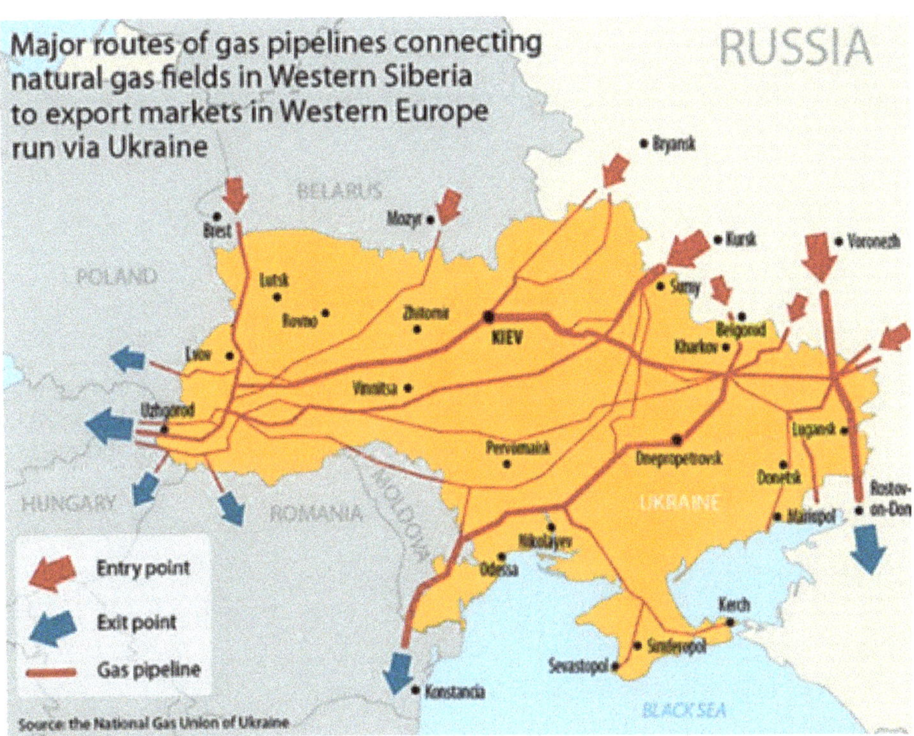

Faltering gas price negotiations and mutual accusations of non-payment and theft had led to a *circa* 3-week standoff and likewise, between Turkmenistan and Iran during the winter of 2007.[9]

---

8 President Joe Biden starts to lift sanctions on Venezuela, *The Economist*, https://www.economist.com/the-americas/2022/11/27/president-joe-biden-starts-to-lift-sanctions-on-venezuela.

9 Turkmenistan and Iran started cooperation in the fields of energy while all of Iran's sectors, including the energy sector, had been under sanctions from the West, especially the U.S. Under a Turkmenistan-Iran natural gas agreement, gas was transported through pipelines from Turkmenistan to Iran. However, the gas was halted when the Tehran administration did not pay for the natural gas that had been imported from Turkmenistan.

These Turkmenistan-Iran and Russia-Ukraine conflicts demonstrate that it is necessary to find an effective way to prevent or manage cross-border disputes in order to mitigate the high economic, political, and human costs associated with cross-border gas supplies. Further, the risks perceived as inherent in cross-border pipelines may increase the cost of financing these projects. Higher financing costs can seriously impact the delivered cost of the gas (with serious consequences for the producers and consumers of gas at both ends of the pipeline).

Thus, it might be advantageous for Trinidad and Tobago and Venezuela to have bilateral cooperation in many areas (other than natural gas) in order to facilitate the long-term strategy of developing Trinidad as a hub for all cross-border gas. Ultimately, Venezuela would avoid the significant cost of developing infrastructure to monetise its gas, and if Venezuela has an interest in monetising its gas in Trinidad, it will mitigate potential gas supply security risks like restricting (or preventing) the flow of gas through the pipelines between the two countries.

Upstream Commercial Terms. In order to obtain its shareholder's approval, an investor upstream has to be provided with commercial terms that make it attractive for him to invest in the particular prospect. In that regard, the costs in respect of exploration and production or an incremental development can only be properly estimated after the individual prospects have been evaluated. Thus, at the outset, although it is important to know the extent to which the licensing terms offered by Trinidad and Tobago compare with those in benchmark countries, the terms on offer will not pass the litmus test if they do not actually lead to a PSC or facilitate an incremental development.

According to Poten, the economics of incremental projects suggest that 1996-2005 PSC terms with gas price indexing of profit gas splits will not support the sanction of many of the developments required to maintain plateau production in the coming years. Poten also concluded that existing production licence terms would similarly struggle to support many new developments:

> *"Improvements in fiscal terms and gas market accessibility may be required to attract preferential investment in Trinidad and Tobago. This would be particularly the case for incumbents under old PSCs and license terms who have to invest to maintain production".*

Should there now be more flexibility when renegotiating the terms for incremental developments?

Bid Rounds. Notably, there was a poor response to the 2022 Trinidad and Tobago deep-water bid Round, with only four of the seventeen blocks receiving single bids by a consortium of bpTT and Shell. Additionally, Woodside Petroleum (formerly BHP) had shown no interest in the deep-water blocks, having acquired extensive seismic data for Trinidad and Tobago's deep-water.[10] Perhaps the Government should now consider sole sourcing strategies (with significantly more favourable commercial terms) in order to attract other upstream players to invest in these blocks?

Upstream commercial concessions today might prove to be more beneficial in the long term:

(a) It is critical to attract upstream players to evaluate the Nation's resource potential.

(b) The bpTT-Shell consortium might leverage its monopoly status in the current gas shortfall context to obtain generous commercial terms in respect of the four blocks.

(c) Both EOG and British Gas were introduced as upstream players when Amoco had an upstream monopoly.

(d) The timing of the Government's objective to determine the entire gas resource potential of the deep-water blocks might not dovetail with conservative bpTT and Shell perspectives.

Long-Term Challenges. The following considerations are also highly relevant when considering Trinidad and Tobago's competitiveness upstream in the longer term:

(a) In Central and South America and the Caribbean regions, countries including Suriname and Guyana, are promoting their own initiatives.

(b) Projects that utilise existing brownfield infrastructure will have a considerable economic advantage due to the lower capital costs that will be incurred.

(c) Maximising the access to ullage in existing facilities will expand the proportion of new developments that can enjoy this advantage.

---

[10] Curtis Williams, 'Deep Water disappointment as only 4 of 17 blocks picked up', *The Guardian*, Thursday 2nd June 2022, https://guardian.co.tt/business/deep-water-disappointment-as-only-4-of-17-blocks-picked-up-6.2.1501318.aef0bd22eb.

(d) The unit lifting cost of gas in Trinidad is likely to increase because unexploited hydrocarbons are being found in increasingly more difficult locations or challenging geological formations with a smaller size of remaining gas pools that require more complex drilling and completion requirements.

(e) Drilling deeper wells or in frontier blocks that are remote from land will be more costly for gas monetisation, and greater volumes of gas production will be required to improve the economics of a development.

Accordingly, in order to encourage further exploration and production and mitigate risks of potential discoveries in new blocks becoming stranded, in the longer term, the Government will have to be even more creative with new PSC terms that change the risk/reward balance.

Climate Change. Despite advances in alternative forms of energy, as matters stand, in the short term, it is still difficult to contemplate a world without refinery or natural gas-based products. These products are used either directly in the transportation industry or as chemical intermediates for several products in a wide range of industrial sectors, including agriculture and food production.

Thus, in the foreseeable future, fossil fuels (including natural gas) are likely to contribute significantly to world economies, which highly depend upon these hydrocarbon resources. It is prudent, therefore, for Trinidad and Tobago to seek to obtain the best value for its energy resources as quickly as possible (and, where feasible, accelerate cross-border projects with neighbouring Venezuela) while adhering to the Nation's international obligations relating to climate change.[11]

LNG or Petrochemicals? Investors, attracted by the relatively low gas prices in Trinidad and Tobago, were encouraged to construct methanol and ammonia plants locally, as a result of which, Trinidad and Tobago became one of the world's leading exporters of ammonia and methanol. However, the export of petrochemicals without the manufacture of local valued-added downstream derivatives[12] represented another method of exporting the natural gas raw material.

---

[11] The Appendix includes the key issues of the Paris Agreement and the main options for its implementation.

[12] The AUM complex that commenced Urea production in 2009 was an exception to the general rule.

And, since primary petrochemical manufacture alone has never generated significant permanent employment, there was an increasing policy preference for gas utilisation for LNG, where greater upstream gas production provided more Government revenue from upstream taxation.

Should the Government now show policy preferences for any value-added operations that aim to create further downstream derivatives from the existing primary petrochemicals? In any event, it will be helpful to understand why local petrochemical producers did not venture into further downstream petrochemical production in Trinidad and Tobago [13] when local gas prices were more internationally competitive. Was it complacency?

Perhaps, with appropriate incentives, this situation could change, particularly if the long-term prospects for gas are improved (which should not be precluded) as a result of extraterritorial supplies from Venezuela. In 2021. Ineos, one of the world's largest manufacturing companies, completed the purchase of the global Aromatics and Acetyls businesses from bp. Ineos Acetyls is a global player in the production of acetic acid and acetic anhydride (which are both downstream derivatives from methanol). Can Trinidad and Tobago attract Ineos Acetyls?

Smart Policy. As a result of the recent Russia-Ukraine conflict, energy prices have risen sharply, significantly benefiting oil and gas producers. Shell and bp, like all private companies, will be motivated to maximise their shareholder value as quickly as possible. And high gas prices ought to benefit Trinidad and Tobago (if the contractual arrangements for LNG marketing permit the full extent of these prices to flow back to the Nation).

However, should the situation change, low LNG prices might not dovetail with Trinidad and Tobago's interests if prices for petrochemicals, particularly ammonia, are projected to be high, and LNG continues to utilise disproportionately large volumes of gas.

Indeed, from time to time, both ammonia and methanol have outperformed LNG as regards product prices. Even in 2008, a time of high HH prices and a high watermark for netback prices

---

[13] Methylamines are readily produced from ammonia and methanol.

for Atlantic LNG, the weighted average netback price from LNG was $4.79/MMBTU, which was better than methanol ($4.31/MMBTU) but significantly worse than ammonia ($6.37/MMBTU).

The depletion of a significant quantity of reserves for LNG could severely undermine the long-term prospects for gas supplies to Point Lisas (and any attendant value-added downstream options). Thus, the Government would benefit by keeping gas utilisation for LNG under constant review and have flexibility in its policy choices for other types of gas utilisation ("smart policy").

State Control of National Resources. In the House of Commons in Britain in 1956, when the subject of the proposed purchase of Trinidad Leaseholds Limited by Texaco became a matter for national debate, the then Chancellor of the Exchequer, Mr. Harold Mac Millan, in his contribution to that debate, said, among other things:[14]

> *"I felt a sense of regret, even dismay, at the thought that an important asset of this kind, hitherto owned and managed by British interests should pass out of our immediate control."*

Dr Williams, Trinidad and Tobago's first Prime Minister had learnt the lessons of history: he understood the importance of State control of its critical resources and its relationship to a country's destiny. Indeed, after Trinidad and Tobago gained independence in 1962, the purchase of bp's assets in 1969 inevitably led to the national ownership and management of bp's marketing outlets. The Trinidad and Tobago National Petroleum Marketing Company (TTNPMC) came into existence in 1972 as a creature of Parliament and took over BP's local marketing activities. ESSO followed in the same year, then Shell, then Texaco. By December 1976, all the local marketing operations previously owned and operated by multinationals were assigned to the TTNPMC.[15]

Thus, when Dr Williams announced the formation of Trintoc at Chaguaramas on Independence Day in 1974, he said:

---

[14] An extract from the 19th Memorial Eric Williams Lecture, delivered by Professor Kenneth Julien (Professor Emeritus) at the Central Bank of Trinidad and Tobago on 5th June 2005. See Appendix 2.
[15] *Ibid.*

> *"The real question was not whether the oil flowed ..... but to whom did the benefits flow –*
> *U.B.O.T. ..... United British first, Trinidad last. Then the name was changed ...... Shell*
> *first, Trinidad last .... Reverse that, put Trinidad and Tobago first – TRINTOC. Now we*
> *know not only from where the oil flows, but to whom the benefits flow."*

It is therefore noteworthy, as one now reflects upon the evolution of the Trinidad and Tobago energy sector, that Shell and bp have completed the following economic cycle of interests:

> Shell, BP, and Texaco first owned Trinidad and Tobago's oil assets. After they left these shores for economic reasons, their upstream and downstream interests were vested in Petrotrin and TTNPMC, both State entities. Today, with the transformation from oil to natural gas, through vertical integration of the LNG chain, they now control a critical engine for the Trinidad and Tobago economy.

Has the Nation been economically re-colonised? *Plus ça change, plus c'est la même chose?*

This cyclical evolution by Shell and bp reflects a somewhat bizarre transformation of the original Vision for State control and energy independence that was first contemplated for the energy sector. In that regard, one might also reflect (once again) on the words of Dr Williams:

> *"There have been attempts to persuade us that the simplest and easiest thing to do would*
> *be to sit back, export our oil, export our gas, do nothing else and just receive the revenues*
> *derived from such exports and, as it were, lead a life of luxury – at least for some limited*
> *period. This, the Government has completely rejected, for it amounts to putting the entire*
> *nation on the dole. Instead, we have taken what may be the more difficult road, and that*
> *is, accepting the challenge of entering the world of steel, Aluminium, methanol, fertilizer,*
> *and petrochemicals. We have accepted the challenge of using our hydrocarbon resources*
> *in a very definite industrialization process."*

Thus, despite the attraction of gas utilisation for LNG due to the high prices caused by war in Ukraine, it is prudent for the Government to keep gas utilisation projects on the front burner if these projects are more likely to facilitate the longevity of the energy sector. Indeed, policymakers should be warned that *those who fail to learn from history are doomed to repeat it.*"

Finally, from the perspective of stability of the Nation, I believe that the success of the energy sector is an important, but not necessarily dispositive, consideration because politicians set the tone for the transformation of society. If they are (or are perceived to be) motivated by short-term self-interests, the Nation, particularly the youth, might dance to their discordant tune, prioritising their self-interests and desires above those of others, which is a destructive force in society.

Ultimately, a Nation cannot survive or develop without positive and creative inputs from its youth.

Appendix 1: The Paris Agreement and Green Hydrogen for Climate Change

The Paris Agreement. The Paris Agreement was adopted on 12th December 2015 and entered into force on 4th November 2016. The Agreement is now the principal regulatory instrument governing the global response to climate change. As of September 2022, 193 States and the EU, representing over 98% of global greenhouse gas emissions, have ratified or acceded to the Agreement. These States include China and the United States, the countries with the 1st and 2nd largest CO_2 emissions among UNFCCC members. Trinidad and Tobago ratified this Treaty on 22nd February 2018, and it came into force on 24th March 2018.

The goal of the Treaty is to **limit global warming** to well below 2 degrees Celsius, **preferably to 1.5 degrees Celsius,** compared to pre-industrial levels. In order to achieve this long-term temperature goal, countries aim to **reach global peaking of greenhouse gas emissions as soon as possible** to achieve a climate-neutral world by mid-century but before 2050.[16] By 2020, countries were required to submit their plans for climate action, known as **nationally determined contributions (NDCs)**. In their NDCs, countries should communicate actions they will take to **reduce their Greenhouse Gas emissions** in order to reach the goals of the Paris Agreement. Additionally, under an **enhanced transparency framework,** starting in 2024, countries are expected to report transparently the actions taken and progress in climate change mitigation, adaptation measures, and support provided or received.

The Obligations.[17] The Agreement is a global Treaty within the meaning of international law, but not all its provisions are legally binding. In that regard, it relies on transparency rather than legal bindingness to promote accountability, and the parties do not have an obligation to achieve their NDCs to address climate change. Thus, States are allowed to determine their mitigation and

---

[16] United Nations, 'The Paris Agreement', https://unfccc.int/process-and-meetings/the-paris-agreement/the-paris-agreement. In its Biennial Update Report on Climate Change, Trinidad and Tobago has committed to an overall cumulative emissions reduction of 15% by 2030 from its three major emitting sectors; https://unfccc.int/sites/default/files/resource/FIRST_%20BUR_TRINIDAD_AND_TOBAGO.pdf.

[17] Daniel Bodansky, Paris Agreement, United Nations Audiovisual Library of International Law, July 2021, https://legal.un.org/avl/pdf/ha/pa/pa_e.pdf.

adaptation actions, and it establishes an iterative process to promote progressively stronger NDCs over time. It establishes the same basic procedural obligations for all parties but allows countries to self-differentiate their substantive mitigation contributions through their NDCs.

Hydrogen. Hydrogen (H_2) is the simplest element known, and it is the most abundant element in the universe. When hydrogen burns, it generates energy in the form of heat, with water as a by-product. Thus, the energy created from hydrogen generates no atmosphere-warming carbon dioxide, making it one of many potential energy sources that could help reduce carbon emissions and slow global warming. $H_2 \longrightarrow H_2O + \triangle$

Natural gas, primarily methane (CH_4), can be converted into hydrogen (H_2) and carbon monoxide (CO) by a process called steam reformation. Currently, hydrogen is used mostly by industry during oil-refining and synthetic nitrogen fertilizer production, and little is used for energy because it is expensive relative to fossil fuels. Water (H_2O) can also be used to produce hydrogen (H_2) by a process called electrolysis.

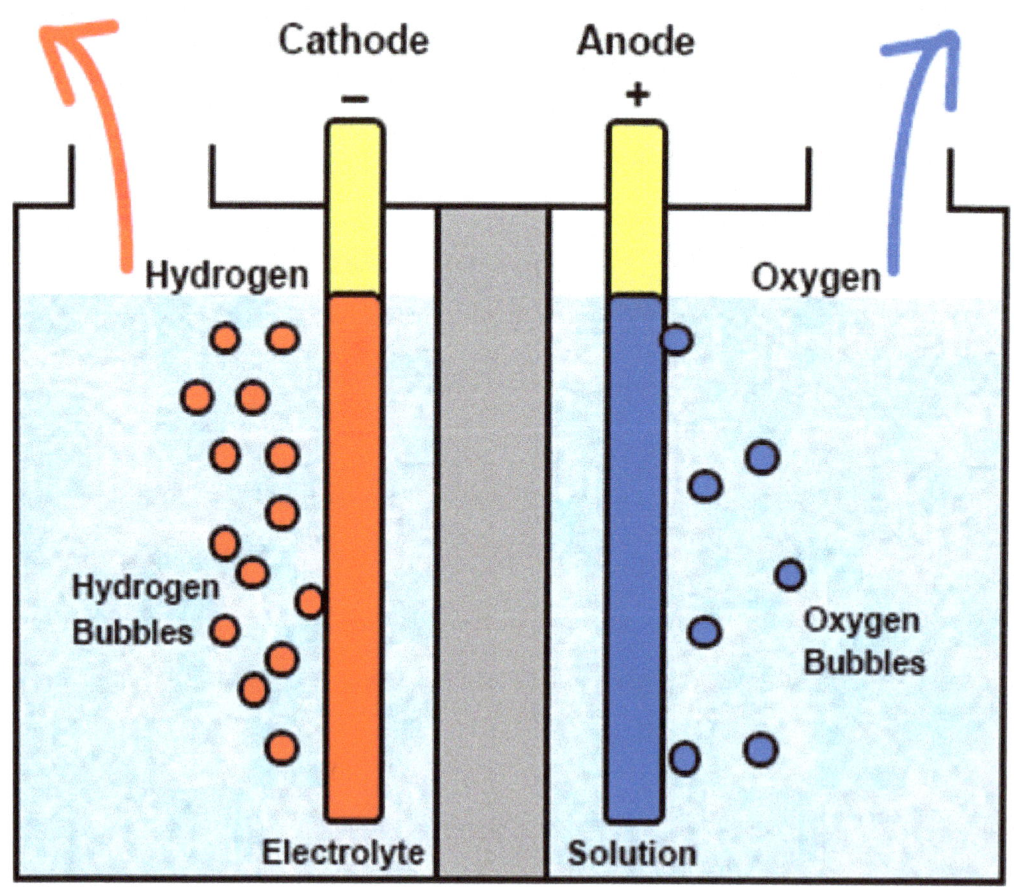

Low-Carbon Hydrogen. The process and energy used determine whether the hydrogen produced is 'low-carbon' or not. If hydrogen is made from natural gas **and** utilises carbon capture, utilisation, and storage (CCUS), it is considered to be low-carbon. The hydrogen produced in this way is referred to as "blue hydrogen." When hydrogen is made from water via electrolysis, if the process is powered by a low-carbon source, such as a renewable energy source like wind, water, or solar photovoltaic systems, there are zero carbon emissions. The hydrogen produced from these processes is commonly referred to as "green hydrogen."

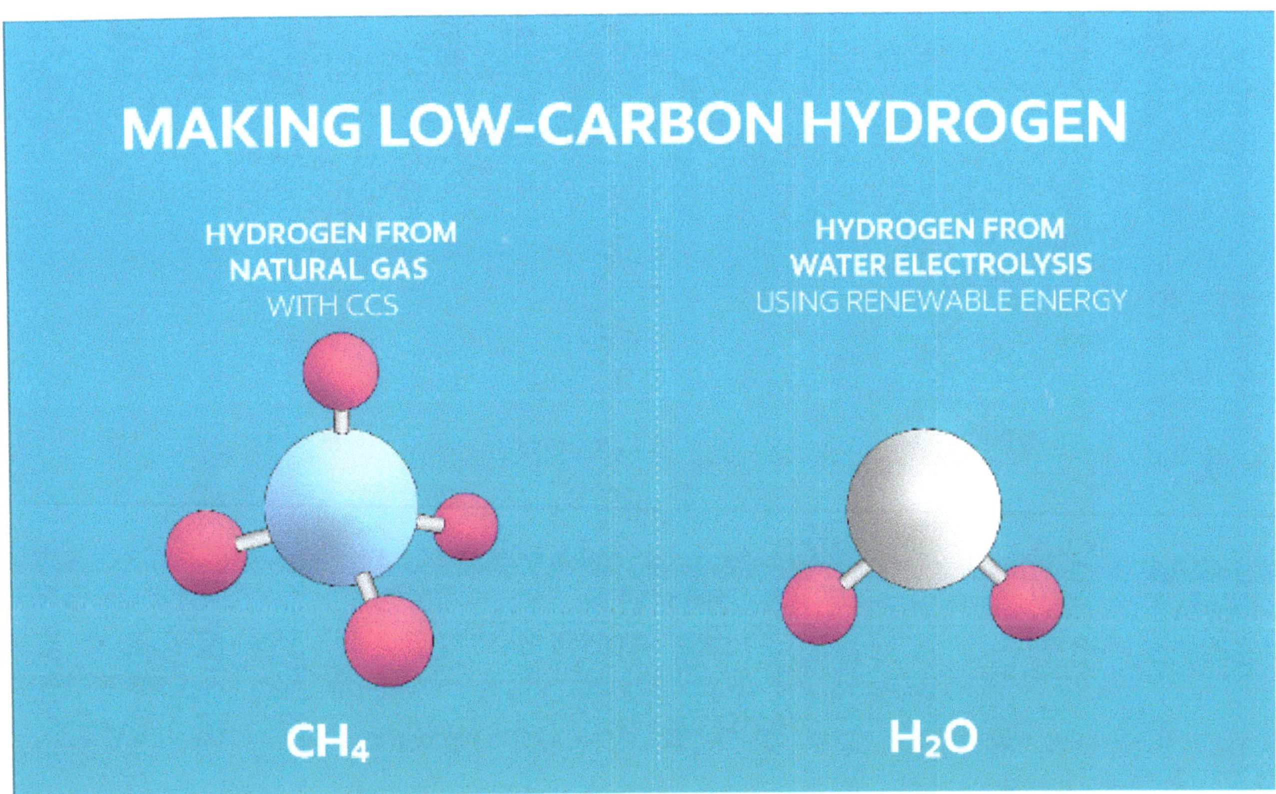

Other Colours of Hydrogen. Today, the vast majority of hydrogen is generated from fossil fuels, particularly from the reforming of natural gas, but also from coal gasification. In the steam reformation process, the hydrogen so produced, without CCUS or capture of the carbon dioxide by-products, is often referred to as "grey hydrogen," contrasting "brown hydrogen" made by coal gasification. "Grey" hydrogen will be converted to "blue" hydrogen if the carbon emissions from the steam reformation process are captured and sequestered underground. Thus, in Trinidad and Tobago, where there is no CCUS, grey hydrogen is used to produce ammonia and methanol.

Whether blue hydrogen meets the threshold of "clean" is a contentious point (discussed later) that depends largely on the ability of carbon capture to make a meaningful reduction in greenhouse gas emissions.

	Colour	Fuel	Process	Products
	Brown/Black	Coal	Steam reforming or gasification	$H_2 + CO_2$ (released)
	White	N/A	Naturally occurring	H_2
	Grey	Natural Gas	Steam reforming	$H_2 + CO_2$ (released)
	Blue	Natural Gas	Steam reforming	$H_2 + CO_2$ (% captured and stored)
	Turquoise	Natural Gas	Pyrolysis	$H_2 + C$ (solid)
	Red	Nuclear Power	Catalytic splitting	$H_2 + O_2$
	Purple/Pink	Nuclear Power	Electrolysis	$H_2 + O_2$
	Yellow	Solar Power	Electrolysis	$H_2 + O_2$
	Green	Renewable Electricity	Electrolysis	$H_2 + O_2$

Costs. Green hydrogen, which has a zero-carbon footprint, also comes at a price. The production cost of hydrogen from natural gas is influenced by a range of technical and economic factors, with gas prices and capital expenditures being the two most important. Fuel costs are the largest cost components, accounting for between 45% and 75% of production costs. As of 2021, according to the U.S. Energy Department,[18] in the U.S., fuel costs of green hydrogen were estimated at about $5 per kilogram compared to $1.50/kg for grey hydrogen.

---

[18] U.S. Department of Energy, Hydrogen and Fuel Cell Technologies Office, '*Hydrogen Shot: An Introduction*', https://www.energy.gov/eere/fuelcells/hydrogen-shot#:~:text=Currently%2C%20hydrogen%20from%20renewable%20energy%20costs%20about%20%245%20per%20kilogram.

The Blue Hydrogen Debate. Natural gas production inevitably results in methane emissions from leaks of methane from the drilling, extraction, and transportation processes. And although methane does not last in the atmosphere as long as carbon dioxide, it is considered to be much more potent as a greenhouse gas.[19] Often, blue hydrogen is described as having zero or low greenhouse gas emissions. However, this is not true because not all carbon dioxide emissions can be captured, and some carbon dioxide is emitted during the production of blue hydrogen. Thus, whether or not blue hydrogen should be pursued is a matter that is hotly debated by some scientists. In that regard, when blue hydrogen production was examined in a peer-reviewed paper,[20] it was found that:

> " … *Far from being low carbon, greenhouse gas emissions from the production of blue hydrogen are quite high, particularly due to the release of fugitive methane. For our default assumptions (3.5% emission rate of methane from natural gas and a 20-year global warming potential), total carbon dioxide equivalent emissions for blue hydrogen are only 9% - 12% less than for grey hydrogen. While carbon dioxide emissions are lower, fugitive methane emissions for blue hydrogen are higher than for grey hydrogen because of an increased use of natural gas to power the carbon capture. Perhaps surprisingly, the greenhouse gas footprint of blue hydrogen is more than 20% greater than burning natural gas or coal for heat and some 60% greater than burning diesel oil for heat, again with our default assumptions. In a sensitivity analysis in which the methane emission rate from natural gas is reduced to a low value of 1.54%, greenhouse gas emissions from blue hydrogen are still greater than from simply burning natural gas, and are only 18% - 25% less than for grey hydrogen. Our analysis assumes that captured carbon dioxide can be stored indefinitely, an optimistic and unproven assumption. Even if true though, the use of blue hydrogen appears difficult to justify on climate grounds.*"

The emissions of methane and carbon dioxide (when using natural gas to produce the heat and high pressure needed for reforming methane and to capture carbon dioxide) could be reduced if

---

[19] Over a 100 year period, one ton of methane is equivalent to 28 to 36 tons of carbon dioxide. IEA, Methane and Climate Change, Methane Tracker 2021, Methane and climate change – Methane Tracker 2021 – Analysis - IEA.

[20] Robert W. Howarth, Mark Z. Jacobson, *How green is blue hydrogen?*, Energy Science & Engineering, 12th August 2021, https://doi.org/10.1002/ese3.956 (Energy Science and Engineering, Volume 9, Issue 10, October 2021, pages 1676 – 1687).

these processes were instead driven by renewable electricity from wind, solar photovoltaic cells, or hydroelectric energy sources. However, the benefit of blue hydrogen has been doubted [21] -

".... this best-case scenario for producing blue hydrogen, using renewable electricity instead of natural gas to power the processes, suggests ... that there really is no role for blue hydrogen in a carbon-free future. Greenhouse gas emissions remain high, and there would also be a substantial consumption of renewable electricity, which represents an opportunity cost. We believe the renewable electricity could be better used by society in other ways, replacing the use of fossil fuels".

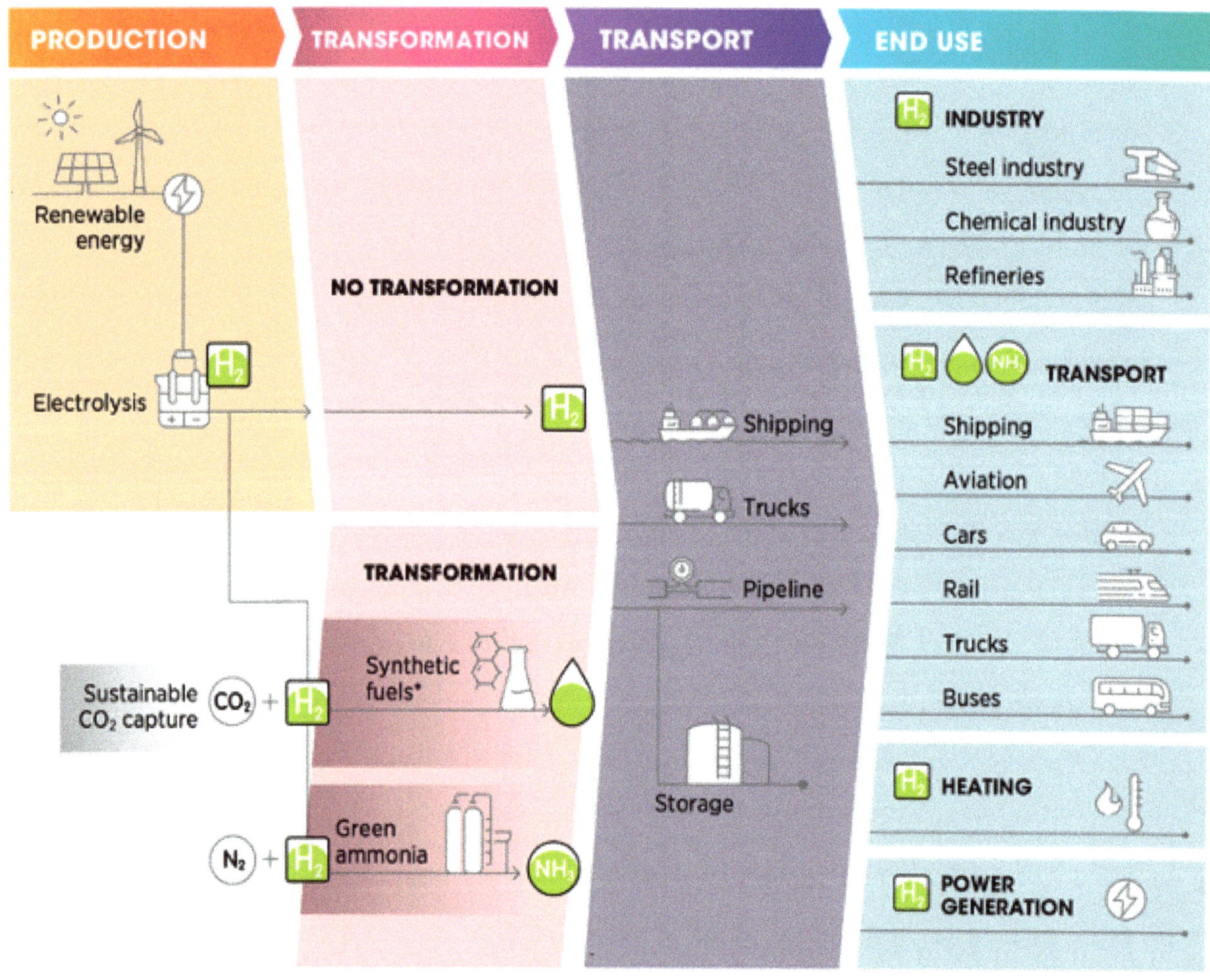

---

[21] Ibid

Some skeptics suggest that green hydrogen is not efficient to be used broadly as an energy source.[22] Critics also argue that the problems with the use of hydrogen as a fuel are fundamental - it takes more energy to produce hydrogen than hydrogen provides when it is converted to useful energy because the process of producing hydrogen, compressing it, and then turning that compressed hydrogen back into electricity or mechanical energy, is grossly inefficient. Thus, according to one analyst,[23] next-generation battery solutions are more attractive:

> *"It's worth putting up with a lot of problems with a battery because for every one joule you put in, you get 90% of it back. That's pretty great ... In producing and storing hydrogen, you get only 37% of the energy back out. So 63% of the energy that you said, is lost. And that's best case."*

Solar PV systems and Wind. Although less than 0.1% of the globally dedicated hydrogen production today comes from water electrolysis, with declining costs for renewable electricity, in particular from solar PV systems and wind, there is growing interest in electrolytic hydrogen.

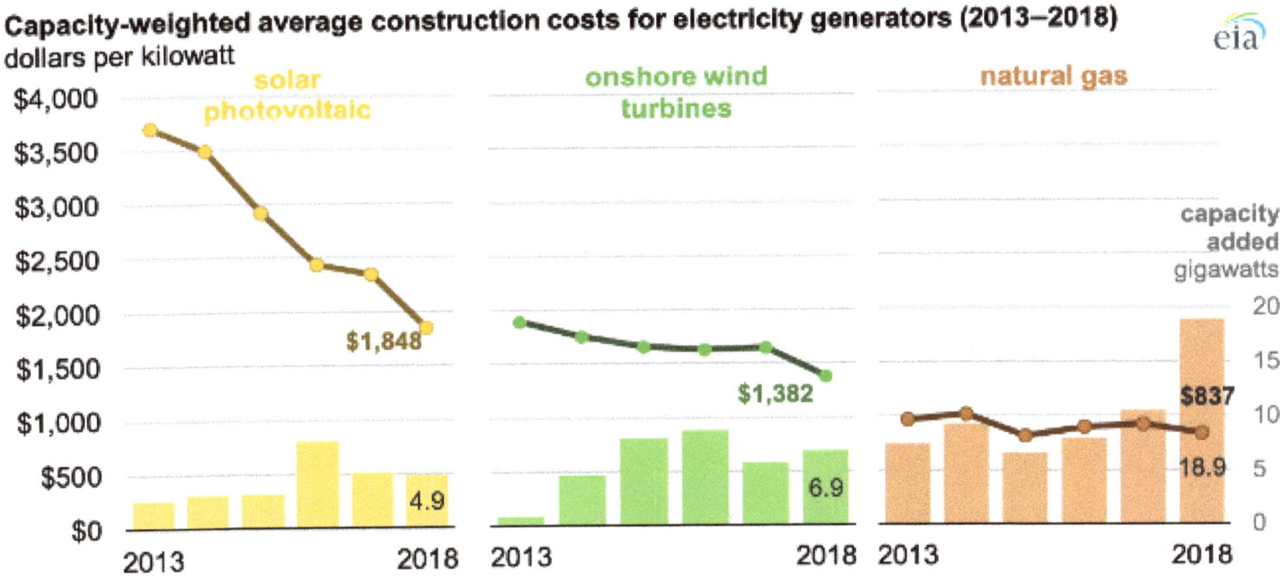

<hr>

22 Robert W. Howarth, Mark Z. Jacobson, *supra*.
23 Catherine Clifford, Clean Energy, 'Hydrogen power is gaining momentum, but critics say it's neither efficient nor green enough', citing comments by Paul Martin, https://www.cnbc.com/2022/01/06/what-is-green-hydrogen-vs-blue-hydrogen-and-why-it-matters.html.

Wind turbines and solar photovoltaic cells convert solar energy flows into electricity. Photovoltaics is the direct conversion of light into electricity at the atomic level by some materials that exhibit a property known as the photoelectric effect. This effect causes these materials to absorb photons of light and release electrons. When these free electrons are captured, an electric current is produced, which can be used to produce electricity. Solar panels are made out of photovoltaic cells.

How a PV Cell Works

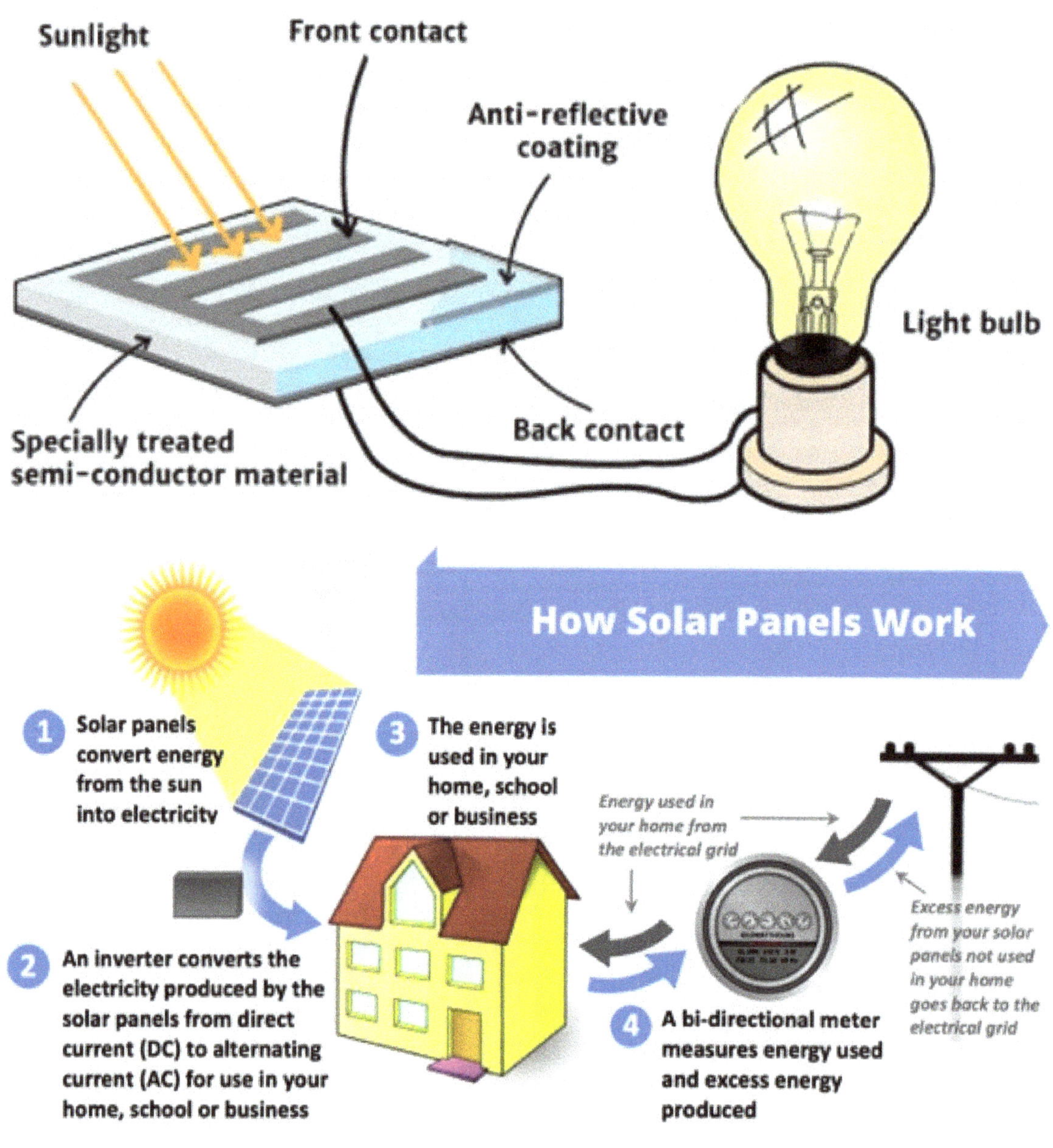

Local Projects. The Ministry of Energy has launched a roadmap, developed by the NEC with support from the Inter-American Development Bank, **to develop a green hydrogen market based on** offshore wind technologies. Preliminary studies indicate that approximately 4MTPY of green hydrogen could be generated, which is more than double the current local demand for the grey hydrogen required by the petrochemical industry. [24] However, in order to convert this vision into reality, it is believed that Trinidad and Tobago will have to invest in the upstream development of the hydrogen value chain, from wind turbines to electrolysis plants, as well as ensure that the downstream infrastructure for green ammonia and methanol is expanded.

Unlike fossil fuels, wind and solar options can only generate electricity when the wind is blowing or the sun is shining, which, despite improved storage solutions over a short timeframe, currently presents engineering challenges because power is generated and consumed simultaneously, with generation varying to keep the system in balance.

In 2022, recently formed NewGen Energy Limited launched the first local project to develop a carbon-neutral/green hydrogen production facility to supply the Tringen facilities at the Point Lisas Industrial Estate. The project aims to produce hydrogen from an industrial electrolysis process powered by carbon-neutral electricity and "green" electricity from renewable sources.[25]

Ultimately, the biggest challenge in the short-term might be the high cost of re-making industries that currently lie at the center of the economy and our lives and the knock-on effects of those increased costs. For example, since green hydrogen generally costs much more to make than less clean types of hydrogen, if it were now used globally for fertiliser production, there would be a severe knock-on effect on food prices.

---

[24] Jonathan Spencer Jones, 'Roadmap for a green hydrogen economy in Trinidad and Tobago', December 7, 2022, https://www.powerengineeringint.com/hydrogen/roadmap-for-a-green-hydrogen-economy-in-trinidad-and-tobago/#:~:text=Trinidad%20and%20Tobago%2C%20the%20first,the%20power%20and%20industrial%20sectors.
[25] Newgen, '*Who we are*', https://newgenenergyltd.com/about-us/.

Appendix 2: Some Defining Moments for the Energy Sector taken from the Eric Williams Memorial Lecture given by Dr Kenneth Julien on 8th June 2005

"Defining Moment – 1963/1964 – No. 1. The Mostofi Commission.

In 1963, one year after Independence, a Commission was established by the Government with the following Terms of Reference:

> *"(1) To examine the present situation and future prospects of the oil industry of Trinidad and Tobago in the context of the economics of the world oil industry;*
>
> *(2) To recommend a legal framework for the oil industry of Trinidad and Tobago which would stimulate the operations of foreign investors while safeguarding the interests of the nation;*
>
> *(3) To make recommendations designed to ensure the greatest possible stability compatible with growth in the industry, including the level of employment."*

The Commission was chaired by Baghair Mostofi with Hamil L. Legall as Secretary.

The recommendations of this Report led to major changes in the legislation that governed petroleum activities in the country and broadened the mandate of the Ministry of Petroleum and Mines which had itself only been established in 1963. Two (2) defining moments rolled into one.

As interesting and far-reaching as the recommendations of this Commission, the questions posed to the Commission by the Government in its official submission are very significant:

> *Have we exploited the natural resources with diligence?*
>
> *Has there been a just division of the proceeds of this natural heritage?*
> *To what extent have we utilized the proceeds of the industry for the betterment of the national as a whole?*

To what extent have we undertaken the training of our nationals for the further exploitation of these resources in the national interest?

Have we taken adequate stock of our international position in all our activities in the industry?

To what extent are laws which may have been appropriate for the operations of the industry under the colonial system compatible with the aims and aspirations and the status of an independent nation pledged to a democratic form of government?

To what extent has there been a plan of development for the industry, and how far has this been coordinated in the Development Plan for the Nation as a whole?

We glean from these questions the hand of Eric Williams as the concept of a national identity for the Energy Sector began to develop. These questions are relevant today as they were forty-two (42) years ago.

Defining Moment No. 2. Acquisition of BP's Assets (1969).

This acquisition of BP's producing assets gave a clear signal that Dr. Williams had begun to take the steps in the pursuit of a policy that will lead to the creation of a National Energy Sector.

The joint venture with Tesoro Corporation to acquire the local producing assets of BP was the first bold step of state ownership in strategic industries. It was prompted by the concern that the closing down of BP's local production would have led to a serious economic and unemployment situation in the St. Patrick area. It tied in with the Eric Williams thinking articulated as early as July 1955 in an address at the University of Woodford Square. He stated:

"There will come occasions when the state may have to take the initiative as an investor, without prejudice to the policy of encouraging and supporting private enterprise, in order to protect and promote the NATIONAL INTEREST".

This bold step of investing in a complex industry such as the petroleum sector, took courage and a strong political will. A key defining moment, less to do with the size of the investment, but more to do with this dramatic move. Notwithstanding this bold move in the late sixties, one gets a sense of Eric Williams struggling to find a clear strategy and the resources to realize his dream and vision of a National Energy Sector.

Three five-year plans were taken to Parliament and while these hinted at this dream, the strategies so far as the Energy Sector was concerned, were vague and not well articulated. In the last five-year development plan (1969-1973), we read a shopping list of petrochemicals, which became the subject of studies:

- Ethylene;
- Thermoplastics;
- Synthetic Fibers;
- Methyl Alcohol; etc.

These were, as expected, all based upon potential products from an oil-based refinery. None of these went beyond these initial studies and none of them had little chance of being realized.

Of greater interest in that document, was the statistic that over the period 1963 to 1968, the natural gas flared was in excess of 50% of the total gas produced. This simple fact, while receiving no comment in the plan, was to trigger the Eric Williams strategy into looking at natural gas for his strategy of industrialization, rather than oil-based products.

Defining Moment No. 3. Natural Gas Discovered off the North Coast (1971).

The story of this discovery and the subsequent commercial production of oil in 1972 is itself a fascinating one, in which Eric Williams played a significant role in persuading AMOCO to have one last try, after a succession of dry holes! …

Defining Moment No. 4. Trinidad and Tobago National Petroleum Marketing Company established by an Act of Parliament – (1972) No. 41.

The purchase of BP's assets in 1969 led inevitably to the need to the national ownership, and management of BP's marketing outlets. The Trinidad And Tobago National Petroleum Marketing Company came into existence in 1972 as a creature of Parliament and took over BP's local marketing activities. ESSO followed in the same year, then Shell, then Texaco. By December 1976, all the local marketing operations previously owned and operated by multinationals were assigned to NATIONAL – TTPMC. The word "National" appearing for the first time, associated with the Energy Sector.

The emergence of a National Energy Sector, which began in the sixties in a tentative manner began to take a more definite shape, catalyzed by discoveries of large reservoirs of natural gas and the dramatic increase in oil prices. Natural gas had crept onto the National Agenda, and Eric Williams' clear message was that priority must be given to our domestic plans of industrial diversification, fueled by natural gas.

At the political level, he found ready allies in Errol Mahabir, Mervyn De Souza, Bunny Padmore, and Patrick Manning. And at the public service and technical level, there was a group of enthusiastic technocrats: Eugene Moore; D. Alleyne; E. Warner; B. Ali; and Ken Julien. All were enthusiastic about following Eric Williams, as he took the country down this bold passage in fulfillment of the vision.

Defining Moment No. 5 – Independence Day 1974. Shell Trinidad became the Trinidad and Tobago Oil Company.

The changes of names of this company were significant:
- UBOT - United British Oilfields of Trinidad
- Shell – Trinidad

And then finally the national identity emerged as **TRINIDAD AND TOBAGO OIL COMPANY – TRINTOC** – August 31st 1974. In his address on that day, Dr Williams said:

"As we proceed to lower the flag of yesterday (the honour falls to the worker with the longest service in Shell, 42 years exactly today) and hoist the flag of today and tomorrow, the flag of the nation as against the flag of an external corporation, as we see the flag, our flag, flying high and riding proud in the breeze, symbolizing the ascent of the nation and the higher destiny of the citizens of Point Fortin, let us say, with pride but yet with humility, we are going well, and may God bless our nation".

Defining Moment No. 6 – 1975. The National Gas Company established in 1975.

The National Gas Company was established in 1975. This company was established for an initial single purpose – to capture the 'so-called free gas' from AMOCO that was being flared and bring it onshore with the expectation that there will be a plan for its use. NGC was born out of this vision – no feasibility study was done or needed. The gas was ours! We had to utilize it. We had to find the money to build the transmission line to bring it onshore.

Defining Moment No. 7. Decision to Invest in ISCOTT – January 17th 1976.

At the time, some of Dr Williams' words said more about his vision of the National Energy Sector rather than about ISCOTT:

"This was the basis of the policy for colonial development, the classic exposition of which was the prohibition on the colonies, especially the mainland colonies in America, expressed by a British Prime Minister, that the colonies were to manufacture not a nail, not a horseshoe. They were to produce raw materials only, which were to be sent to England, to enable downstream manufacturing operations, to provide jobs, to expand.

Our presence here today at Point Lisas testifies to the fundamental changes that have taken place in the world economy and in the economic balance of power. On the one hand sugar has gone; only Cuba and Brazil survive today to recall the so-called glory that was sugar in the ancient colonies. Beet sugar in Europe and America has supplanted us, and Australia has revolutionized the old colonial policy by exploding the fallacy that the sugar industry was not for white people. Point Lisas is the symbol of this fundamental reorientation of the international economy. Sugar cane gives way to wire rods. Sugar has separated us as wire rods will weld us back together.

There have been attempts to persuade us that the simplest and easiest thing to do would be to sit back, export our oil, export our gas, do nothing else and just receive the revenues derived from such exports and as it were, lead a life of luxury – at least for some limited period. This, the Government has completely rejected, for it amounts to putting the entire nation on the dole. Instead, we have taken what may be the more difficult road and that is, accepting the challenge of entering the world of steel, aluminium, methanol, fertilizer, petrochemicals. We have accepted the challenge of using our hydrocarbon resources in a very definite industrialization process."

The die was cast.

Defining Moment No. 8: Joint Venture with WR Grace - TRINGEN was born. 1977

At the opening of the Tringen Plant, [Dr. Williams'] words can say much more than I can about this rapid emergence of the National Energy Sector:

"Eighteen years ago, the supply of natural gas to the plant was negotiated between WR Grace and a transnational oil company, with little or no reference to the Government. For the Tringen Project, negotiations took place between Tringen and the Government who owns the gas and the transmission facilities which bring the gas to the plant. Eighteen years ago, the average price negotiated for the gas was 27 cents per thousand cubic feet. Today, the gas is being sold by the Government to Tringen at $1.68 cents per thousand cubic feet, some six times more."

Defining Moment No. 9. Signing of Contract – Fertrin.

This was a joint venture between the Government of the Republic of Trinidad and Tobago (GORTT) and AMOCO. At the ceremony marking this occasion, Dr. Williams made the point that it was the first time that the Government was entering into a joint venture (51 per cent GORTT, 49 per cent AMOCO), "with a major oil-producing company – one active in oil and gas production in Trinidad and Tobago." He did not disclose publicly the agonizing and private debate that went on before responding to this proposal from AMOCO.

Firstly, [there was] the question of AMOCO's motives. AMOCO had committed to establishing a refinery once a certain level of oil production had been reached. This proposal for a joint venture provided AMOCO with an opportunity to remove that commitment which did not at the time fit into their global plan. Secondly, a joint venture with a large multinational was contrary to his thinking that the sardines of this world have to keep their distance from the sharks.

Firstly, he summed it up in graphic language, which I will convey into words more befitting this audience:

> *"Professor, whenever a developing country gets into bed with a large multinational, it is very, very, likely that it will lose its VIRGINITY. It is a risk we have to take but ensure that we do not give up any of our SOVEREIGNTY."*

The national identity and aspirations must be preserved at all costs! Did Trinidad and Tobago lose its virginity? Did we preserve our sovereignty? Another story, another time.

These words have guided us even to this date, as we continue to treat with the several industrial giants in our midst, hopefully with greater confidence. An academic question: how many times can we lose our virginity!

Defining Moment No. 10. The Start of Operations of ISCOTT.

On December 5th 1980, Dr. the Honourable Eric Williams, then Prime Minister and Minister of Finance, presented the 1981 Budget. One extract of that presentation reads as follows:

> *"The decision of the Government to move boldly into the field of industry based on the use of our energy resources has been the subject of discussion, debate, criticisms, and at times outright hostility generated both internally and externally. Those decisions have been translated into one producing unit, ISCOTT, and by the middle of 1981, another additional to the productive sector of our economy, Fertrin. In parallel with these developments and in support of them, have been the establishment of a modern industrial estate, 1,500 acres in extent at Point Lisas, a modern Port and Harbour facility to accommodate vessels of 50,000 tonnes dead weight and a Power Plant to meet the demand of proposed industries at Point Lisas and the country as a whole. Within a matter of four years, the gas consumption dedicated to these industrial developments will have increased from almost zero to some 500 million cubic feet per day. A use has been found for our natural gas, the alternative to which would have been flaring it and burning it, or saving it for export to some large metropolitan country with a thirst for cheap energy."*

This was very likely, his last major public statement.

On March 29th 1981, he passed away."